# TRANSFORMATIONS OF ELECTRICITY IN NINETEENTH-CENTURY LITERATURE AND SCIENCE

*To my husband James,*
*for his wisdom and his unwavering support and faith in me.*

# Transformations of Electricity in Nineteenth-Century Literature and Science

STELLA PRATT-SMITH

Routledge
Taylor & Francis Group

LONDON AND NEW YORK

First published 2016 by Ashgate Publishing

Published 2016 by Routledge
2 Park Square, Milton Park, Abingdon, Oxon OX14 4RN
711 Third Avenue, New York, NY 10017, USA

*Routledge is an imprint of the Taylor & Francis Group, an informa business*

**British Library Cataloguing in Publication Data**
A catalogue record for this book is available from the British Library

**The Library of Congress has cataloged the printed edition as follows:**
Pratt-Smith, Stella, 1965-
   Transformations of electricity in nineteenth-century literature and science / by Stella Pratt-Smith.
      pages cm
   Includes bibliographical references and index.
   ISBN 978-1-4724-1940-8 (hardcover)
1   English literature--19th century--History and criticism. 2.  Science in literature. 3.  Electricity in literature. 4.  Literature and science--Great Britain--History--19th century.  I. Title.

   PR468.S34P73 2016
   820.9'356--dc23
                                                                              2015029929

ISBN 9781472419408 (hbk)

# Contents

# List of Figures

# Preface

This book is about a series of extraordinary transformations, ones that converted electricity from an unexplained natural wonder to a core driving force of modern technology. It is also about how many of the nineteenth century's most creative thinkers in literary and scientific spheres tried to get to grips, through words and symbols, with one of Nature's fundamental energies. The fascination of the topic has long been apparent to scholars and non-specialists alike. Electricity is commonly understood as a unique phenomenon and it was recognised as such in the nineteenth century more than at any other time in history. Electricity's extensive and contradictory list of features made it not only invisible and particularly difficult to investigate but also sensational, terrifying, beautiful, dangerous, seductive and, just as crucially, useful. The creativity prompted by that rare combination of qualities spanned the disciplines, bringing about a nexus of interests as extensive as it was inclusive.

Much of the material examined here is drawn from beyond the literary canon. The scope is, admittedly, ambitious. While well-known works are considered, also compared here are responses from scientific experts and amateur enthusiasts, short fiction and commentaries by periodical columnists, novels that have long since lapsed into obscurity, and reactions from contemporary readers. The book brings together this wealth of works, writers, forums and audiences in order to illuminate their connections and disjunctions, and to offer amongst other things a relatively wide-ranging account of nineteenth-century authorship and reading about electricity. The goal is, as a whole, to give a sense of just how energetic, lithe and elaborate interactions were during the period, between writers, reading publics, disciplines and contexts.

Nineteenth-century writers certainly had a variety of reasons for writing about electricity or referring to it in their works; this book seeks to investigate those and to prompt further questions. What did writers think the mention of electricity added to their writing? What purposes were actually performed by its representation? And where do these intentions and purposes meet, if at all? I ask also how these functions varied from one form of writing to another, what the features were that appeared most prominently in scientific, popular and fiction writings on the topic and, very much as part of these other queries, how did contemporary readers respond? What value did they think representations of electricity had, in terms of communicating or subverting scientific or other knowledge? These are the questions that prompted the project in the first place, and it has been a privilege to have the opportunity to explore them in such depth.

The research for this volume was made possible by a grant from the Arts and Humanities Research Council, to whom I am hugely grateful. I owe further thanks for their valuable advice and assistance to my doctoral supervisor Michael Whitworth at Merton College, Oxford, to the wider Oxford English faculty, and to my doctoral examiners Sally Shuttleworth and Gowan Dawson. I have been tremendously fortunate to benefit also from the support of several other colleagues and friends, who contributed to the research at various stages and encouraged me along the way. Alice Jenkins and Barri Gold, I would like to mention as particular and continual inspirations throughout the four years of research, not just as gifted scholars but also as exceptional people and friends. I would also like to thank many other people for their help and support at various points, including Amanda Caleb, Bernard Lightman, Andrew Scott Mangham, John Holmes, James Mussell, Stefano Evangelista, Robert Douglas-Fairhurst, Nick Shrimpton, Geoffrey Cantor, Graeme Gooday, Frank James, Mark McCartney, Claire McKechnie, Tim Chiou, Jessica Hancock and Ruth Schuldiner. I am deeply grateful to the University of Glasgow for the outstanding undergraduate and masters teaching I had, especially from Kirstie Blair, Andrew Radford, Christine Ferguson, Alison Chapman, Willy Maley, Richard Cronin, and Susan Hill.

I am indebted, too, to the associations whose conferences I have attended and been involved with during my research, especially the British Society for Literature and Science, the British Association for Victorian Studies, the Midlands Interdisciplinary Studies Seminar and the Interdisciplinary Nineteenth-Century Studies society. Each of these forums allowed me to benefit from an array of thought-provoking and stimulating papers by participants from all over the world. I am also hugely thankful to my undergraduate students at Oxford, whose insight and energy helped invigorate my own efforts and has always made teaching a true privilege.

On a personal level, I am eternally grateful to my wonderful mother, Jean, who has always been willing to talk through and apply her remarkable intelligence to even my quirkiest ideas. I can hardly express my gratitude to her, or to my late father Leighton Pratt, an extraordinary, deeply sensitive and gallant man, who died in 2008 on the very day I was accepted to Oxford to start the research for this book. It was a tragic irony that he would have understood and appreciated more than most, and which he would have felt keenly for me. He is ever in my thoughts. Finally, this book is dedicated to my husband, James, whose fortitude, strength, love and friendship have taught me to attempt much more than I would ever have thought possible. He has taught me always to see things through, despite the difficulties and, as a result, to value myself as well as others. With him by my side, I hope that, in due course, this book will be the first of many.

# Chapter 1
# Creative Sparks

'There is perhaps no branch of experimental philosophy, which is received by persons of all ages with greater pleasure than Electricity', declared Henry Minchin Noad in 1844.[1] The phenomenon was, he suggested, 'calculated to arrest the attention' and had the capacity to become more powerfully 'fixed on the mind' than any other. By the time he wrote, electricity had emerged as a distinct focus of study but, far from demystifying the phenomenon, scientific investigation continued to enhance electricity's widespread and gripping appeal. As Noad also points out, electricity was perceived to have 'secret and hidden influence' connected to 'the most sublime and awful' agencies of nature. In this and a variety of other ways discussed here, writers of all kinds revealed electricity's troubling allure and its fundamentally problematic nature.

How electricity was written about shaped not just public perceptions of the phenomenon, but also the development of scientific understandings about it and its potential applications. Literary responses to electricity were amalgamations of scientific, literary and cultural concerns that related directly to the phenomenon and its study. Of course, many nineteenth-century sciences, such as biology, geology, botany and anthropology, also sought to elucidate the historical and material basis of the natural world and its relationship to human existence. Investigations of electricity were different though, because they also confronted a phenomenon that appeared to be essentially intangible. Writing about electricity was especially problematic, for it meant describing something that had never been visualised or depicted before and, rather than being classified neatly as 'scientific', it was open to speculation by all. At the same time, while electricity could be approached by empirical means, actually understanding its properties and representing them meant embracing previously inconceivable levels of abstraction and complexity. In that sense, examining electricity was simultaneously a part of and beyond the dominant visual culture of the nineteenth century that has become so widely recognised by scholars in recent years.[2] Even as electricity was explored

---

[1]  Henry Minchin Noad, *Lectures on Electricity: Comprising Galvanism, Magnetism, Electro-Magnetism, Magneto- and Thermo-Electricity* (London: George Knight and Sons, 1844), 3. Noad's popular books about electricity are discussed further in Chapter 3.

[2]  Kate Flint, *The Victorians and the Visual Imagination* (Cambridge: Cambridge University Press, 2000); see also Carol T. Christ and John O. Jordan, eds, *Victorian Literature and the Victorian Visual Imagination* (Berkeley: University of California Press, 1995); Lindsay Smith, *Victorian Photography, Painting and Poetry: The Enigma of Visibility in Ruskin, Morris and the Pre-Raphaelites* (Cambridge: Cambridge University

in terms of physical matter by scientists, just as frequently, it was perceived to distance man from the otherwise 'natural' world of direct, known experience. Investigations of electricity were closely affiliated to contemporary technological applications, developments that referred primarily to imagined futures and an essentially transient present, rather than a documentable past. The combination of these features made electricity a rich and varied imaginative resource but, as the writings I consider indicate, it also made literary responses to the phenomenon inherently unstable.

## Electricity in the Nineteenth Century

The period investigated here begins in 1831 with Michael Faraday's discovery of electromagnetism and concludes with the first public supply of electricity in 1881, a time when electricity gained entirely new and, arguably, unprecedented significance in the public consciousness.[3] Of course, people had long been aware of electricity, in witnessing the natural phenomena of lightning or static shocks. What made electricity distinct in the nineteenth century was its emergence as a unified scientific concept, with uses, meanings and implications never previously anticipated. The way in which electricity and experimentation with it was represented involved confronting a range of obstacles, and not just scientific ones. Neither did nineteenth-century understandings of electricity develop in a straightforward or linear fashion. While narrative connections between writings often arose, a more marked tendency overall was the *lack* of teleological and epistemological progression between texts, readings and authorships in the period. Despite attempts in the 1840s and 1850s by the likes of Michael Faraday, James Clerk Maxwell, William Sturgeon and Arthur Smee, to dispel common myths about electricity, the latter continued to re-emerge 30 and even 40 years later, in such fictions as Benjamin Lumley's *Another World* (1873) or anonymous short stories like 'Doctor Beroni's Secret' (1884), both of which are examined later in this book. Parallels and progressions did exist between types of authorship, genres and content; however, just as many divergences emerged. It was the combination of the two that shaped the period's understandings and beliefs about electricity.

Between 1830 and 1880, such genres as the novel and the short story were still very much in the process of becoming established, so writings of the period can only ever be loosely assigned to such specific categories. Writings about electricity particularly defied epistemological and disciplinary boundaries, employing

Press, 1995); Jonathan Smith, *Charles Darwin and Victorian Visual Culture* (Cambridge: Cambridge University Press, 2006).

3    Ian McNeil, ed., *An Encyclopaedia of the History of Technology* (London: Routledge, 1990), 369. The first public supply of electricity was offered in Godalming, Surrey, in the form of street lighting. The Electric Lighting Act was introduced in 1882 and amended in 1888, after which electric street-lighting proliferated.

features that belonged to a range of literary forms, with continual and not always logical exchanges of technical, literary and cultural precepts. Scientific research on electricity drew on fictional concepts and emerging principles of physics were brought into poetic responses, just as non-fiction incorporated narrative or anecdotal techniques, and real scientific developments were integrated and explored in fictional works. Rather than understanding these different forms of literature on the basis of a singular 'genre-level meaning', I suggest that interchangeability was an integral and even characteristic aspect of the writings.[4]

Electricity's emergence as a simultaneously literary and scientific phenomenon implicitly contests the opposition of the two spheres and supports, instead, the 'one culture' model.[5] This was even more so between 1830 and 1880, when logical, empirical and consistent understandings about electricity were still being established. The precarious status of knowledge about electricity's properties and processes disrupted, in itself, any possibility of conventional scientific and literary distinctions. Definitions of the term 'literary' frequently rely on relatively vague notions of creativity and artistic merit, considerations that writings about science tend, just as often, to subvert. Assumptions about epistemological characteristics are misleading here, too, not least because they construct stereotyped portrayals of scientific and literary practices, as well as their subsequent development. As Laura Otis notes, 'to understand how nineteenth-century people thought about communications, it is essential to read the works of scientists and novelists in parallel. Although they lived and worked quite differently, they faced the same challenge to communicate and answered it with cultural knowledge and creativity'.[6] My approach to writings about electricity treats them as frequently dissimilar in purpose but, nonetheless, similarly motivated by their authors' equal desires to investigate, to communicate and to explore creative possibilities. As interdisciplinary research, the present work welcomes the pluralism of nineteenth-century readerships and seeks out literatures beyond the established canon, to interrogate conceptions of literature and extend our understandings beyond just 'the best which has been thought and said in the world'.[7] Recognising literature's culturally and socially constructed nature has created a broader appreciation of

---

[4] Dallas Liddle, *The Dynamics of Genre: Journalism and the Practice of Literature in Mid-Victorian Britain* (Charlottesville: University of Virginia Press, 2009), 154.

[5] Gillian Beer, *Darwin's Plots: Evolutionary Narrative in Darwin, George Eliot and Nineteenth-Century Fiction* (London: Routledge and Kegan Paul, 1983); Sally Shuttleworth, *George Eliot and Nineteenth-Century Science: The Make-Believe of a Beginning* (Cambridge: Cambridge University Press, 1984); George Levine, ed., *One Culture: Essays in Science and Literature* (Madison: University of Wisconsin Press, 1987).

[6] Laura Otis, *Networking: Communicating with Bodies and Machines in the Nineteenth Century* (Ann Arbor: University of Michigan Press, 2001), 6.

[7] Matthew Arnold, *Culture and Anarchy: An Essay in Social and Political Criticism* (London: Smith, Elder and Co., 1869), viii.

'non-literary' language.[8] Yet the goal of learning how stories and poems make meaning is still described as 'literary' by historians of literature and science.[9] Analysing writings about electricity requires being vigilant yet embracing, too, the slippery and extendable nature of the term 'literary', exploring not just how texts make meaning but also the kinds of meaning they make, and remaining keenly aware throughout of how those meanings are influenced by different authorial perspectives, forms of writing and intentions.

My research does not suggest the existence of a hierarchy of knowledge, whereby scientific understandings of electricity might be considered innately more or less valuable than their depiction by popular or fiction writers. In the nineteenth century as much as now, narrative and fictional elements existed in a variety of writings, specialist or not, so that as Ralph O'Connor suggests, 'rhetorical tropes and aesthetic forms did not merely decorate the science presented, but helped to construct it'.[10] Fictionality was a core feature of scientific conceptualising and representational techniques were understood as vital aspects of scientific progress, albeit sometimes awkward ones. Explanations of electricity were forms of literary response crucially shaped by a symbiotic relationship between content, purpose and reading practices. In both fiction and non-fiction, writers interpreted, transformed and created new associations with electricity and, in that sense, the conceptualisation and representation of electricity was inseparable from literary, social and material contexts. Exchanges between literature, science and society were relatively fluid for, as Patricia Fara points out, 'both literature and science are mutually shaped by each other and by the communities which generate them'.[11] To emphasise this inseparability is not to 'reduce the technical content of the sciences to a nexus of social interests', for which sociologists of scientific knowledge have been criticised in the past.[12] It does reveal, however, the unique combination of technical content and social interests upon which writings about electricity relied. As Gillian Beer suggests, 'most major scientific theories rebuff common sense. They call on evidence beyond the reach of the senses and overturn the observable world. They disturb assumed relationships and shift what has been substantial into metaphor'.[13] Many of the writings considered in my research reveal exactly this amalgamation of the substantial, the fictional and the metaphorical. Scientific

---

[8]    Elena Semino, *Metaphor in Discourse* (Cambridge: Cambridge University Press, 2008), 42.

[9]    Laura Otis, 'Science Surveys and Histories of Literature: Reflections on an Uneasy Kinship', *Isis*, 101:3 (September 2010), 571.

[10]    Ralph O'Connor, *The Earth on Show: Fossils and the Poetics of Popular Science, 1802–1856* (Chicago: University of Chicago Press, 2007), 227.

[11]    Patricia Fara, *Sympathetic Attractions: Magnetic Practices, Beliefs, and Symbolism in Eighteenth-Century England* (Princeton: Princeton University Press, 1996), 173–4.

[12]    Myles W. Jackson, 'A Cultural History of Victorian Physical Science and Technology', *The Historical Journal*, 50 (2007), 264.

[13]    Beer (1983), 3.

discoveries and theoretical understandings informed some writings but, just as often, misunderstandings, accidents, inventions and translations led to creative innovation and still farther-reaching propositions.

Electricity was frequently contemplated in the nineteenth century according to its materiality or spirituality too. As Alice Jenkins contends, 'to equate the physical sciences in this period with an unambiguously materialist outlook is gravely to misread the evidence'.[14] Certainly, the persistence of a confusing 'heterogeneity' of electrical theory created considerable space for uncertainty, contradiction and speculation.[15] One way of penetrating that confusion is to approach it as a jostling array of multi-faceted perspectives, comprehensible ultimately as an assemblage of diverse participants striving to represent electricity in any way they could. Many of the writings examined in this volume propose associations between electricity and a range of other contemporary interests that have fallen foul since of increasingly rigorous definitions of the term 'scientific'. But then a whole unsteady gamut of electrical theories, hypotheses and laws was always fundamental to portrayals, speculations and explorations of electricity. Indeed, the relationship between literature and science has long been recognised as one of 'interchange rather than origins and transformation rather than translation', wherein 'scientific discourses overlap, but unstably'.[16] Electricity, in particular, represented an 'immaterial, conceptual space' that demanded new forms of 'spatial imagination', as well as an array of authorial purposes, intellectual exchanges, publication forums and genres.[17]

In considering how writers conceptualised ideas of electricity and experimentation with it, the 'diffusion model' of science and society, whereby science influences literature but literature does not appear to influence science, tends to fall by the wayside. Instead, as Barri Gold notes in relation to nineteenth-century literary engagements with science, 'literature has often, perhaps always, influenced science, especially in the delicate, early stages of a scientific development, before a phenomenon has been named or a hypothesis adequately articulated'.[18] This distinctive, lively and dynamic 'conversation' is particularly evident in writings about electricity.[19] Indeed, engaging with ideas about energy and electricity is made possible only by the frequent boundary crossing that we

---

[14]    Alice Jenkins, *Space and the 'March of Mind': Literature and the Physical Sciences in Britain, 1815–1850* (Oxford: Oxford University Press, 2007), 6.

[15]    Martin Willis, *Mesmerists, Monsters, and Machines: Science Fiction and the Cultures of Science in the Nineteenth Century* (Kent, OH: Kent State University Press, 2006), 71.

[16]    Gillian Beer, *Open Fields: Science in Cultural Encounter*, 2nd edn (Oxford: Oxford University Press, 2006), 173.

[17]    Jenkins (2007), 3.

[18]    Barri J. Gold, *ThermoPoetics: Energy in Victorian Literature and Science* (Cambridge, MA: MIT Press, 2010), 15.

[19]    Gold (2010), 28.

now describe as interdisciplinarity, a venture that revisits our understandings of these intricate relationships and enables us to disclose their further, intriguing epistemological affinities.

A primary feature of this book, as mentioned previously, is its concentration on writings beyond the canon, and the valuable evidence of diversity they offer, in terms of nineteenth-century literatures, publication forums and readerships. Periodical writings represent not just important but essential elements in the broader catalogue of the period's 'literary' works. For some time, scholars felt that the 'sheer bulk and range of the Victorian press' made it 'so unwieldy as to defy systematic and general study'.[20] However, the 'electronic harvest' recognised since, of digitised archives and enhanced research tools, has led to a dramatic expansion of scholarship on both periodicals and popular writing.[21] The realisation has transpired that, in the nineteenth century, 'periodicals rather than books provided the main means of dissemination'.[22] Although there is still much work to be done before popular science writing in periodicals is acknowledged to the same extent as other literary forms of the period, the value of periodical texts and the diversity of genres and epistemologies they represent is better understood now than ever before. The publication market beyond the reviews and costly journals offers a wealth of writings on electricity, and ones that stray beyond the editorial and stylistic constraints of publications that were weightier, as well as less frequently published. By drawing on the relatively cheap and broadly circulated periodicals, supplied to a range of audiences in terms of class, educational levels and ideologies, fresh insights can be identified between science, writing and reading.

Fiction about electricity did not appear only in novels but in a selection of short stories, many of which were published in monthly publications. A particularly striking change in the nineteenth century was the growth in the circulation of 'penny fiction-based weeklies catering to a relatively uncultivated audience' of

---

[20]    Joanne Shattock and Michael Wolff, eds, *The Victorian Periodical Press: Samplings and Soundings* (Leicester: Leicester University Press, 1982), xiii.

[21]    James Secord, Review: 'The Electronic Harvest', *The British Journal for the History of Science*, 38:4 (Dec., 2005), 463; Susan Sheets-Pyenson, 'Popular Science Periodicals in Paris and London: The Emergence of a Low Scientific Culture, 1820–1875', *Annals of Science*, 42:6 (November 1985), 549–72; John Christie and Sally Shuttleworth, eds, *Nature Transfigured: Science and Literature, 1700–1900* (Manchester: Manchester University Press, 1989); Ruth Barton, 'Just before Nature: The Purposes of Science and the Purposes of Popularization in Some English Popular Journals of the 1860s', *Annals of Science* 55 (1998), 18–26; G.N. Cantor, Gowan Dawson, Graeme Gooday, Richard Noakes, Sally Shuttleworth and Jonathan R. Topham, eds, *Science in the Nineteenth-Century Periodical: Reading the Magazine of Nature* (Cambridge: Cambridge University Press, 2004).

[22]    Geoffrey Cantor, Gowan Dawson, Richard Noakes, Sally Shuttleworth and Jonathan R. Topham, 'Introduction', in Louise Henson, Geoffrey Cantor, Gowan Dawson, Richard Noakes, Sally Shuttleworth and Jonathan R. Topham, eds, *Culture and Science in the Nineteenth-Century Media* (Aldershot: Ashgate, 2004), xviii.

readers, who had only minimal levels of formal education.[23] Penny-weekly writers tended to refer to electricity only minimally, in the form of exhibition reviews or descriptions of past scientific figures, rather than engaging more directly with ideas about electricity.[24] As frequently as elsewhere, the idea of electricity is limited to its deployment as an adjective, for example, in describing the powerful emotions of a young heroine flattered by a stranger, where the 'flattery flies to the heart as swiftly as electricity along the wire'.[25] Such metaphorical responses and their functions are considered at various points throughout this volume, as popular explorations of electricity, contemporary experimentation with it, and its wider ramifications.

At this point, we need to pause and query what constitutes the 'popular' anyway? Writings about electricity appear in an eclectic array of forums, in portraying scientific activity, figures and knowledge beyond the laboratory. Many of the writings explored here are what might be termed 'popular', a word that that touches immediately on the acknowledged yet still painfully limited nature of terms available for scholarly discussions of science writings, participants and practices. Terms such as 'popularisation' and 'popular science' carry awkward political and ideological baggage, including inaccurate assumptions about class and cultural boundaries, and derogatory top-down models of the 'popular' as a watered-down and poorly understood form of 'real' science.[26] In reality, the products of élite scientific culture were not merely used by the 'lower orders' of the scientific hierarchy; they were also transformed, through richly creative processes that produced different yet equally valid forms of science. A mixture of 'popular' and scientifically 'élite' writings about electricity are discussed in this book, which demonstrate the true diversity of authorial purposes that existed in the nineteenth century.

The limitations, associations and assumptions of how we discuss 'popular' discourses are well-noted.[27] The obstacles faced by popular science, in achieving

---

[23]   Gowan Dawson, Richard Noakes and Jonathan Topham, 'Introduction', in Cantor et al. (2004), 18.

[24]   For example, 'A Day at an Electro-Plate Factory', *Penny Magazine*, 13:806 (26 October 1844), 417–24; Albert Smith, 'The Adventures of Mr. Ledbury', *Bentley's Miscellany*, 12 (July 1842), 556; Edwin F. Roberts, 'The Road to Happiness in Six Steps', *Reynolds's Miscellany of Romance, General Literature, Science, and Art*, 1:11 (23 September 1848), 169 to 1:16 (28 October 1848), 249 (in six parts); John V. Bridgeman, 'Shocks', *Train: A First-Class Magazine*, 4:20 (August 1857), 111–15.

[25]   'Florence May – A Love Story', *Chambers's Journal of Popular Literature, Science and Arts*, 2 (January 1854), 26.

[26]   Roger Cooter and Stephen Pumfrey, 'Separate Spheres and Public Places: Reflections on the History of Science Popularization and Science in Popular Culture', *History of Science*, 32 (September 1994), 237–67.

[27]   James A. Secord, 'Knowledge in Transit', *Isis*, 95 (2004), 654–72; Bernard Lightman, *Victorian Popularizers of Science: Designing Nature for New Audiences* (Chicago: University of Chicago Press, 2007); Jonathan R. Topham, 'Rethinking the History

credibility or validity, lay partly in its non-specialist origins. Yet many Victorian popularisers of science were also relatively expert, and gave entirely accurate scientific explications, despite the 'popular' nature of their writings and lectures. The difference between 'specialist' forms of science and their more popular versions seems to arise largely from the accessibility of the latter. Indeed, the meaning of the term 'popular' has shifted from 'belonging to the people' to 'presenting knowledge in generally accessible ways', indicating that the ownership of knowledge changed, from originating with 'the people' to being received by them. Certainly, there were also 'owners' of early nineteenth-century science who strove to make knowledge more accessible and many scientists were aware of the benefits of this movement. Indeed, after the low sales of Mary Somerville's *On the Connexion of the Physical Sciences* (1834), the author frequently assured her editors that she would make her writings more understandable to wider audiences.[28]

As the nineteenth century progressed, more accessible forms of science were often taken less seriously as scientific discourse, while attitudes towards 'popular science' were shaped increasingly by changing publication climates and readerships for scientific writings. The inclusion of science writings by new publications made scientific knowledge something that was no longer simply produced by élite specialists and consumed by passive, non-scientific under-classes. There was a new dynamic at work, whereby producers and interpreters of scientific knowledge catered as much, if not more, to readers' wishes as to scientific hierarchies – so much so that, between 1781 and 1840, the Royal Society's 'monopoly' was overthrown by the foundation of some two dozen specialist scientific societies.[29] It is hardly surprising, then, that so many new forms of science and access to it were perceived by scientific authorities as threatening. It is no coincidence that the anxieties prompted by wider information dissemination and mass literacy prompted such publications as the *Popular Science Monthly* to voice reservations about the 'reading crowd'.[30] Dismay over the wider distribution of news and the emotions it might prompt in readers meant that 'in England, many writers were equally fascinated and troubled by the notion of a contagious spread of emotions and ideas amongst crowds and newspaper publics'.[31] The metaphor of disease is particularly apt for the period, conveying as it does the ostensibly positive possibilities of change, alongside such threats as social unrest, lawlessness and

---

of Science Popularization/Popular Science', in *Popularizing Science and Technology in the European Periphery, 1800–2000*, ed. F. Papanelopoulou, A. Nieto-Galàn and E. Perdiguero (Aldershot: Ashgate, 2009).

[28]    James Secord, ed., *Collected Works of Mary Somerville*, vol. 1 (Bristol: Thoemmes Continuum, 2004), 63.

[29]    Richard Yeo, *Defining Science: William Whewell, Natural Knowledge and Public Debate in Early Victorian Britain* (Cambridge: Cambridge University Press, 1993), 33.

[30]    Caroline Sumpter, 'The Cheap Press and the Reading Crowd: Visualizing Mass Culture and Modernity, 1838–1910', *Media History*, 12:3 (December 2006), 233.

[31]    Sumpter (2006), 240.

epidemics. As the writings examined here demonstrate, alongside excitement and hope, the scientific fascination with electricity prompted equally disturbing levels of fear, anxiety and hostility.

## Electricity and the Canon

'Historians have long urged', Anne Secord suggests, 'that a fuller understanding of nineteenth-century science will be gained if we broaden our sense of what constitutes scientific activity'.[32] The novels discussed in this chapter illustrate just such a broadening of what is meant by 'scientific activity', straying as they do into spheres as diverse as the metaphorical association between electricity and emotion, the relationship of electricity to time, space and evolution, and the visualisation of future medical treatments. As today's wealth of scholarship on works beyond the canon demonstrates, canonicity is no longer a prerequisite for investigating works with other valuable features.[33] It is interesting to note, nonetheless, the relative ubiquity of metaphorical references to electricity by leading Victorian novelists, including Dickens, the Brontës, George Eliot, James Froude, Thackeray, and Wilkie Collins. Further literary responses to electricity, represented by William Harrison Ainsworth's *Auriol, or the Elixir of Life* (1850), Edward Bulwer-Lytton's *A Strange Story* (1862) and *The Coming Race* (1871), and Benjamin Lumley's *Another World, or Fragments from the Star City of Montalluyah* (1873), illustrate the complexity of interactions between nineteenth-century concepts of electricity and other contemporary developments.

As a literary form, the novel may appear to have become increasingly 'elevated' from other forms of writing in the nineteenth century, as well as distanced from scientific writings.[34] However, we can also note that the 'privileged moment of individualization' represented by the novel form is largely an illusion.[35] Novels were created in contexts that were powerfully influenced by scientific ideas, processes of popularisation and periodical writings. Technically, the nineteenth-

---

[32]   Anne Secord, 'Pressed into Service: Specimens, Space, and Seeing in Botanical Practice', in *Geographies of Nineteenth-Century Science*, ed. David N. Livingstone and Charles W.J. Withers (Chicago: University of Chicago Press, 2011), 286.

[33]   For example, Ruth Bernard Yeazell, *Sex, Politics, and Science in the Nineteenth-Century Novel* (Baltimore: Johns Hopkins University Press, 1990); Alan Rauch, 'Science in the Popular Novel: Jane Webb Loudon's *The Mummy!*', in *Useful Knowledge: the Victorians, Morality, and the March of the Intellect* (Durham, NC: Duke University Press, 2001), 60–95; Robin Anne Reid, *Women in Science Fiction and Fantasy: Overviews* (Westport: Greenwood Publishing Group, 2009).

[34]   William B. Warner, 'The Elevation of the Novel in England: Hegemony and Literary History', *ELH*, 59:3 (Autumn 1992), 577–96.

[35]   Michel Foucault, 'What is an Author?', tr. Donald F. Bouchard, in *Language, Counter-Memory, Practice*, ed. Donald F. Bouchard and Sherry Simon (Ithaca: Cornell University Press, 1977), 124.

century novelist's response to ideas about electricity was quite different to that of the periodical contributor or scientific collaborator; however, novelists exist within societies and periods and, while their responses are not necessarily representative, they still offer interpretations of scientific ideas that reveal associations and ideas within the period, and that invariably connect with a range of readerships. The nature of fiction makes it difficult to measure its precise or tangible contribution to scientific practice, but we can consider how novels allowed contemporary writers and readers to respond to prevailing concepts about electricity.

References to electricity, their meanings and purposes, may seem inseparable from the genre in which they occur; however, works by Ainsworth, Bulwer-Lytton and Lumley often teeter precariously between being fictions about science or 'science fiction'. Rather than getting bogged down here in anachronistic distinctions of literary merit, we can start by approaching the novels through nineteenth-century views of scientific fiction. The term 'science fiction' appears to have originated in 1851 with the fiction author William Wilson (OED), for whom the primary value of the genre lay in 'creating interest where, unhappily, science alone might fail'.[36] For Wilson, fictions about science elicited quite different responses in readers than instructions or explorations of science; yet he does not view fiction as an introductory form of science. As he observes, his contemporaries felt science fiction had the capacity to convey 'the truths of Science' and that these could be 'interwoven with a pleasing story which may itself be poetic and *true*' (author's emphasis).[37] Fictional frameworks were not considered to detract from scientific merit; they both accompanied and fortified scientific insights. Wilson's usage of the term 'science fiction' differs from its twentieth-century use; he views its role as a vehicle for conveying accurate facts and scenarios about science, whereas it is the futuristic, fantasist and technological elements of the genre that later come to the forefront. Nevertheless, as David Seed suggests, 'exploration lies at the heart of SF'.[38] The history of electricity's industrial and scientific development is also one of experimentation, bringing writings about it into close alignment with both conceptual and literary experiment.

The century's new discoveries, Wilson remarks, have thrown 'deeply into shade the old romances and fanciful legends'; for him, the magnetic needle 'has more magic about its *reality*, than the wildest creations of child-fiction and legend have in their *ideality*' (author's emphases).[39] He refers to Hans Oersted's discovery in 1820 that not only is a magnetic needle deflected by an electric current, but the live electric wire is also deflected in a magnetic field. Breakthroughs such as this,

---

[36]    William Wilson, *A Little Earnest Book on a Great Old Subject* (London: Darton and Co., 1851), 137.

[37]    Wilson (1851), 139.

[38]    David Seed, ed., 'Introduction: Approaching Science Fiction', in *A Companion to Science Fiction* (Blackwell Publishing, 2005), Blackwell Reference online, 22 August 2011. Please note that page numbers are not indicated in this edition.

[39]    Wilson (1851), 143.

in understanding electromagnetism, provided contemporary fiction authors with especially rich ground for intermingled scientific and literary meanings. From the 1820s onwards, researchers investigated, described and explained electricity, but fiction authors were free from the constraints of accuracy. This meant writings were taken less seriously by both science and literature, but it also allowed authors to explore the phenomenon in more creative forms, from the realistic to the absurd and occult. Indeed, as Nathaniel Hawthorne comments, the 1840s was a time when 'the comparatively recent discovery of electricity and other kindred mysteries of Nature seemed to open paths into the region of miracle'.[40] Scientific interest in electricity fuelled literary speculation and vice versa; although scientific romance appeared to prevail, it was promoted by the developments in electrical science, rather than diminished by or opposed to them.[41] Fiction authors expressed real contemporary interests and anxieties, and engaged with a number of concurrent and competing themes beyond electricity or science. In that sense, speculations about electricity were no more divorced from scientific speculation than from the rest of the period's dominant interests.

### Electrical Metaphors in the Novel

By the beginning of the nineteenth century, parallels were already well-established between electricity and particular associations, images and scenarios. Existing analogies between electricity and strong emotion, for example, allowed novelists to make allusions, to which readers could immediately relate. As Laura Otis suggests, metaphor allows both scientists and novelists to convey their ideas to a variety of readers; indeed, they 'seize upon any ready analogies culture has to offer' and 'forge their own metaphors, which then enter the cultural store'.[42] My consideration of responses to electricity in novels works outwards from the metaphorical references authors make. At the same time, I seek to show how they contributed to the 'cultural store', as well as reinforcing and extending widespread perceptions of the phenomenon.

Charles Dickens's deployment of electrical metaphors in the novel begins with *The Old Curiosity Shop* (1840).[43] He describes as 'electrical' the effect of

---

[40] Nathaniel Hawthorne, 'A Birth-Mark', *The Pioneer* (March 1843), reprinted in *Mosses from an Old Manse* (London, 1846).

[41] See Nicola Bown, Carolyn Burdett and Pamela Thurschwell, *The Victorian Supernatural* (Cambridge: Cambridge University Press, 2004), 1. The term 'scientific romance' has been attributed to H.G. Wells, who used it in reference to his own work in the 1920s; see George Slusser, 'The Origins of Science Fiction', in *A Companion to Science Fiction*, ed. David Seed (Blackwell Publishing, 2005), Blackwell Reference online.

[42] Otis (2001), 6.

[43] The first figurative use of the term 'electrical' is attributed to Laurence Sterne, *The Life of Tristram Shandy*, II, 19 (1760) (OED).

cooking smells upon a crowd coming in from the rain to eat, conveying effects that are simultaneously sudden, invisible and reviving by alignment with the same apparent features of electricity.[44] Dickens refers largely to atmospheric electricity and the telegraph in his novels.[45] However, he also uses the metaphor of electricity to accentuate visible effects, firstly, in *Martin Chuzzlewit* (1844), when Mr Pecksniff starts back 'as if he had received the charge of an electric battery' and in *Dombey and Son* (1848), when the response to a dinner-table story is described as 'like an electric spark'.[46] The speed and power of electricity makes it appropriate as a metaphor for emotional shocks and for fear.[47] In *Dombey and Son*, electricity becomes the metaphor for a particularly interesting type of fear. When Carker is on the run, creeping through the streets at night, we are told that 'some other terror came upon him quite removed from this of being pursued, suddenly, like an electric shock'.[48] The reference to electricity signals a transformation of his fear from the ordinary and justified to the 'visionary', 'unintelligible' and 'inexplicable'. It is mentioned at the moment when a presence flies out of the darkness, 'associated with a trembling of the ground, – a rush and sweep of something through the air, like Death upon the wing'. Carker shrinks back, 'as if to let the thing go by'; yet as soon as it is gone, he knows that 'it never had been there' and he is left bewildered, with a sense of 'startling horror'. Carker's fear is a premonition of his death struck by a train, but it is accentuated by what is described as its 'electric' quality and the indefinable phenomenon he experiences has the invisibility, swiftness and danger of electricity, as well as that of modern technology. Theorists of metaphor indicate that 'no claims have been made in the literature as to which metaphors are particularly important to fear' and that 'the conceptualization of fear is conventionally based on cold, not heat'.[49] However, the use of electricity as a metaphor for fear by Dickens and several of the other authors discussed here indicates that it was important. The electrical metaphor for fear has a physiological element, too, in the rather inelegantly termed sensation of 'piloerection' (*cutis anserine*), more commonly known as the feeling of one's hair standing on end with fear. It is now known that the sensation is the result of

---

[44]    Charles Dickens, *The Old Curiosity Shop: A Tale* (London: Chapman and Hall, 1841), 189. The editions of novels referenced here are, wherever possible, those closest to the original publication dates.

[45]    Dickens refers to atmospheric electricity and the electric telegraph in *Dombey and Son* (1848), *Hard Times* (1854) and *Our Mutual Friend* (1865).

[46]    Charles Dickens, *The Life and Adventures of Martin Chuzzlewit*, vol. 1 (London: Chapman and Hall, 1844), 598; *Dombey and Son* (London: Bradbury and Evans, 1848), 365.

[47]    The phenomenon is discussed from the perspective of 'psychic' shock in Jill Mathus, *Shock, Memory and the Unconscious in Victorian Fiction* (Cambridge: Cambridge University Press, 2009).

[48]    Dickens (1848), 545.

[49]    Anatol Stefanowitsch and Stefan Thomas Gries, *Corpus-Based Approaches to Metaphor and Metonymy* (Berlin: Walter de Gruyter, 2006), 93; Zoltán Kövecses, *Metaphor in Culture: Universality and Variation* (Cambridge: Cambridge University Press, 2005), 28.

adrenaline surging from the adrenal gland to the sympathetic nervous system, but this was not discovered until the last decade of the nineteenth century.[50] Before then, the speed of lightning combined with known experiences of electricity on nerves and muscles indicated that it might be the source of physical feeling.[51] The physical sensation of fear being so akin to that of electrical shock meant that, before further explanations were devised, the analogy was very suitable.

Moments of emotional disturbance also provide the basis of the Brontës' use of electrical metaphors. When Emily Brontë's Heathcliff is startled by a creaking oak, he feels it 'like an electric shock', and when Anne Brontë's Mrs Graham is startled by Markham's shadow, it 'gave her an electric start'.[52] Electricity denotes a disturbing nervousness and mental restlessness in characters, which also epitomised life's unpredictability. One of the most iconic and memorable images of traumatic romance in Charlotte Brontë's *Jane Eyre* (1847) is of a tree torn asunder by lightning. In Brontë's later novel *Shirley* (1849), Caroline Helstone asks 'what is that electricity they speak of, whose changes make us well or ill; whose lack or excess blasts; whose even balance revives?'[53] Caroline soliloquises over the influences that exist 'about us in the atmosphere, that keep playing over our nerves ... now a sweet note, and now a wail'.[54] Like Dickens's portrayal of Carker's experience, Brontë's depiction of electricity gives it a level of agency and personification, one that is spectral or at least ephemeral in character. In *Villette* (1853), the characterisation becomes distinctly diabolical. When Lucy Snowe is battling for creative inspiration with a French composition, she experiences the 'rushing past of an unseen stream of electricity', which she interprets as the herald of a stirring 'irrational demon ... strangely alive'.[55] The similarity of Lucy's experience to Carker's in *Dombey and Son* suggests that, despite increasing scientific knowledge, in the public imagination, electricity continued to be seen as a disturbing, unfathomable and ghostly phenomenon.

In Zoltán Kövecses's recent study of metaphor and emotion, he proposes that emotions may be 'constructed' using a combination of metaphor, culture and physiology, and he points out the metaphorical association between emotions and electricity with the example of 'an electrifying experience'.[56] Electricity was indeed used by nineteenth-century novelists to describe experiences but it also provided a

---

[50]   The discovery of adrenaline is most frequently attributed to George Oliver and Edward Schäfer at University College, London in 1895; see Walter Sneader, *Drug Discovery: A History* (Chichester: John Wiley, 2005), 155. Other papers on the subject were also produced around the same time in Poland and the United States.

[51]   Sidney Ochs, *A History of Nerve Functions: From Animal Spirits to Molecular Mechanisms* (Cambridge: Cambridge University Press 2004), 108–29.

[52]   Emily Brontë, *Wuthering Heights* (London: Thomas Cautley Newby, 1847), 54; Anne Brontë, *The Tenant of Wildfell Hall* (London: T.C. Newby, 1848), 130.

[53]   Charlotte Brontë, *Shirley* (1849; repr. London: John Murray, 1929), 433.

[54]   Brontë (1849).

[55]   Charlotte Brontë, *Villette* (London: Smith, Elder and Co., 1853), 70.

[56]   Kövecses (2005), 83.

metaphor for the essence of characters' personalities. In Thackeray's *The History of Pendennis* (1849), he adopts perceptions of the phenomenon as a metaphor for superficial emotion. When Laura Bell describes Blanche Amory's friendship towards her as 'a very sudden attachment', Blanche replies with characteristic effusiveness that 'all attachments are so. It is electricity – spontaneity. It is instantaneous. I knew I should love you from the moment I saw you. Do you not feel it yourself?'[57] Although Blanche uses the idea of electricity to profess the immediacy of her affections, in fact the alignment with electricity emphasises her staccato style of speech, as well as her impetuous and inconstant nature.

The analogy between powerful emotions and electricity offers George Eliot a way to suggest further physical and metaphysical connections. In Eliot's *Scenes of Clerical Life* (1858), for example, a 'deep bass note' rings out from a harpsichord struck by Oswald, and Eliot writes that 'the vibration rushed through Caterina like an electric shock'.[58] It is a revelatory moment for Caterina, when we are told that 'it seemed as if at that instant a new soul were entering into her, and filling her with a deeper, more significant life'.[59] Again, the physical and emotional aspects of the experience are connected by means of comparison to electricity. Later in the novel, Janet experiences a similar epiphany when she remembers Mr Tryan's sympathy towards her and 'the thought was like an electric shock'.[60] The reference compares, too, to the experience of Noel Vanstone's housekeeper Mrs Lecount in Wilkie Collins's *No Name* (1862), who reaches a moment of unexpected mental clarity through the effect of electricity. Mrs Lecount wakes in the night with her 'head whirling as if she had lost her senses' and yet 'with electric suddenness, her mind pieced together its scattered multitude of thoughts, and put them before her plainly under one intelligible form'.[61] The metaphors exploit many of the contemporary speculations and demonstrations of electricity's properties discussed in other chapters here. As James Delbourgo suggests, the use of electrical metaphors to describe spiritual awakenings 'resonated because of the material culture of electrical demonstration'.[62] Metaphor provided a way of portraying fictional experiences and, simultaneously, making observations of the phenomenon itself. Electricity's innate speed and abruptness suggest, for example, the tense liveliness of Collins's characters, whose eyes are 'hardly ever in repose' or 'strike through' others 'with an electric suddenness', whose abruptness 'fairly took away' the

---

[57]    William Makepeace Thackeray, *The History of Pendennis* (Leipzig: Bernard Tauchitz, 1849), 19.

[58]    George Eliot, *Scenes of Clerical Life* (Edinburgh and London: William Blackwood and Sons, 1858), 33.

[59]    Eliot (1858).

[60]    Eliot (1858), 257.

[61]    Wilkie Collins, *No Name*, ed. Mark Ford (1862; repr. London: Penguin, 1994), 308.

[62]    James Delbourgo, *A Most Amazing Scene of Wonders: Electricity and Enlightenment in Early America* (Cambridge, MA: Harvard University Press, 2006), 134.

breath, or who are unusually aware of 'nervous influences'.[63] Metaphors drew on shared perceptions of electricity's inherently powerful yet capricious nature, and supplied further associations by aligning them with key attributes of memorable fictional characters.

The metaphor of love as electricity operates on a somewhat different level to the examples considered so far. Describing love in electrical terms is one of several metaphors pointed out in the seminal work *Metaphors We Live By* (1980) by Lakoff and Johnson, together with the interpretation of love as a patient, madness, magic or war.[64] Like Lakoff and Johnson, Kövecses defines as a 'physical force' the metaphorical use of electricity to describe touch; however, he emphasises its association with 'sexual *magnetism*' (Kövecses's emphasis) by listing it in the table 'Metaphors and Metonymies of Lust'.[65] The metaphorical association with sexuality is evident in nineteenth-century fiction, albeit less explicitly so, within the self-imposed censorship of contemporary discourse. In *Adam Bede* (1859), Eliot compares the integrated relationship of electricity to air with that of love and enthusiasm, claiming that 'our love is inwrought in our enthusiasm as electricity is inwrought in the air'.[66] After the mid-nineteenth century, the pejorative overtones of fanaticism with the 'enthusiasm' she refers to were diminishing towards associations with intense, spiritual and artistic emotion.[67] In *Daniel Deronda* (1876), the electrical metaphor allowed Eliot to describe both new and old associations simultaneously, to express the profound communication between Mirah and Daniel. As Mirah watches Daniel clasp Mordecai's hand for the first time, the narrator observes that the movement 'seemed part of the flash from Mordecai's eyes, and passed through Mirah like an electric shock'.[68] Eliot employs electricity to denote the prophetic power and invisible bonds of heritage, a kinship that is both communicated and felt in physical terms.

Eliot appears to suggest her own century's greater scientific understanding of electricity, when she reflects on the previous century as a time when man 'had not had the slightest notion of that electric discharge by means of which they had all wagged their tongues mistakenly'.[69] However, she also endows it with mysterious qualities, particularly when she refers to Deronda 'touching the electric chain of

---

[63]   Collins (1994), 8, 270; *Man and Wife* (Leipzig: Bernhard Tauchnitz, 1870), 122; *The Moonstone* (New York: Harper and Brothers, 1868), 202.

[64]   George Lakoff and Mark Johnson, *Metaphors We Live By* (Chicago: University of Chicago Press, 1980), 49; George Lakoff and Mark Johnson, *Philosophy in the Flesh: The Embodied Mind and its Challenge to Western Thought* (New York: Basic Books, 1999), 72.

[65]   Zoltán Kövecses, *Metaphor and Emotion: Language, Culture, and Body in Human Feeling* (Cambridge: Cambridge University Press, 2003), 31.

[66]   George Eliot, *Adam Bede*, vol. 2 (Leipzig: Bernhard Tauchnitz, 1859), 89.

[67]   Susie I. Tucker, *Enthusiasm: A Study in Semantic Change* (Cambridge: Cambridge University Press, 1972), 27.

[68]   George Eliot, *Daniel Deronda*, vol. 4 (London: William Blackwood and Sons, 1876), 246.

[69]   Eliot (1876), vol. 3, 129–30.

his own ancestry'.[70] The same metaphor is used by James Froude in *The Nemesis of Faith* (1849), when the protagonist Markham Sutherland exclaims

> To be an author – to make my thoughts the law of other minds! – to form a link, however humble, a real living link, in the electric chain which conducts the light of the ages! Oh! how my heart burns at the very hope.[71]

Eliot knew Froude from her time editing the *Westminster Review* in the 1850s, and it has been suggested that her English translation of *Leben Jesu* (1846) prompted the religious doubts expressed in his partly autobiographical novel.[72] Down the 'ages' of the intervening decades, Froude's metaphor connects to *Daniel Deronda*'s concern with literary influence, heritage and connectivity, creating precisely the 'living link' to which he refers. Electricity as a phenomenon provides a way in which to figure the invisible yet 'vital' power of metaphysical connectedness.

Froude's protagonist also states, 'I use magnetic illustrations, not because I think the mind magnetic, but because magnetic comparisons are the nearest we have, and the laws are exactly parallel'.[73] The same would appear to be the case with electricity. The metaphors employed by authors adopt earlier understandings of vitality and electrical fluid, which were still undergoing development. As Nicholas Roe suggests, they 'served as a helpfully ambiguous go-between, a material *numen* infusing matter with a less tangible life force'.[74] The function provides the basis of an ongoing analogy between romantic love and electricity, as a physically energising force. We can see this in Wilkie Collins's *Armadale* (1866), when Mr Bashwood meets Miss Gwilt again and his 'motive power' is 'annihilated by the electric shock of her touch and her look'. The electrical analogy prompts transformations on several levels. Despite Bashwood having already become a 'worn-out old creature who had not sung since his childhood', he is transported to a 'seventh heaven of fatuous happiness' and, like an early electrical battery, he is depicted afterwards literally humming with delight.[75]

The metaphorical responses considered here suggest that nineteenth-century authors viewed electricity with an unusual combination of fascination, delight and wariness. Yet, as George Slusser suggests, 'the mention of science is quasi-absent from mainstream novels'.[76] The examples discussed here indicate that, when mainstream novelists did engage with scientific ideas, they often did so only briefly

---

[70]    Eliot (1876), vol. 4, 196.

[71]    James Froude, *The Nemesis of Faith* (London: J. Chapman, 1849), 44.

[72]    Owen Chadwick, *An Ecclesiastical History of England: The Victorian Church* (Oxford: Oxford University Press, 1966), 532.

[73]    Froude (1849), 134.

[74]    Nicholas Roe, *Samuel Taylor Coleridge and the Sciences of Life* (Oxford: Oxford University Press, 2001), 8.

[75]    Wilkie Collins, *Armadale*, vol. 2 (London: Smith, Elder and Co., 1869), 606.

[76]    Slusser, in Seed (2005).

or by means of metaphor, rather than direct reference. Rather than investigating the internal detail of scientific technicalities, they explore the peripheries where scientific concepts meet human experience. They portray what individuals experience when they encounter what *feels* like electricity and they attempt to depict the relationship between the experience and similarly indefinable, invisible phenomena, such as clarity of mind or high emotion. While references to electricity may appear to be minimal or indirect in the novels discussed, in fact, they provide critical ways of conveying aspects of experience that would be hard to communicate otherwise.

**Auras, Elixirs and Enchantment**

From the 1830s onwards, fiction authors added their own impressions of electrical exploration, as well as their hopes and fears for its future influence. This is particularly evident in William Harrison Ainsworth's *Auriol, or the Elixir of Life* (1850).[77] Although the novel was reviewed as 'one of the best of the author's works', its literary merit remains debatable.[78] However, it is one of the earliest novels to engage at length with ideas about electricity and it offers a number of ideas and associations that indicate wider perceptions of the phenomenon.

The title's reference to the 'elixir of life' offers a gamut of associations relating to electricity, an intricate network of links between electricity, ancient mysticism, materiality, wealth and literary style. 'Auriol' is also the name of the novel's hero but it does not exist as a word; instead, it is an approximation and amalgamation of several similar words, which suggest the novel's hybrid nature. It is closest in spelling and pronunciation to three words: 'ariole', meaning a soothsayer or diviner; 'aureole', which refers to a surrounding golden halo or corona, from the Latin for gold (*aurum*) on which the periodic symbol 'Au' is based; and, finally, 'ariel', the spirit who serves Prospero in Shakespeare's *The Tempest*.[79] Like the metaphorical references already discussed, the words have associations with mystery and mysticism, creating expectations that are at least partially fulfilled by the novel's opening.[80] Alice Jenkins suggests that 'literature and science studies

---

[77] The edition referenced here is William Harrison Ainsworth, *Auriol, or the Elixir of Life* (1850; London: George Routledge and Sons, 1890). The novel was serialised in the *New Monthly Magazine and Humorist*, 74:295 (July 1845), 421–31 to 76:301 (January 1846), 109–12. Ainsworth owned and edited the *New Monthly Magazine* from 1845 until 1870, as well as several other periodicals. He wrote around 40 popular historical romances, including *Guy Fawkes* (1840), *The Tower of London* (1840) and *Old Saint Paul's* (1841).

[78] Review: 'Auriol, and other Tales', *Critic*, 9:233 (15 December 1850), 591.

[79] 'Ariel' was also the name given to a moon of Uranus discovered by William Lassell on 24 October 1851. The two moons previously discovered by William Herschel on 11 January 1787 were named 'Titania' and 'Oberon', and the planet's 27 moons were subsequently named after characters in the works of Shakespeare and Alexander Pope.

[80] Scientific associations with 'Darcy' do not appear to be relevant, as 'Darcy's Law' of fluid dynamics and hydrology was not devised until after Ainsworth's novel; see Henry

needs to pay attention not only to the challenge from the new but to the desirability for many purposes of the old'.[81] The allure of 'the old' is evident from the outset of Auriol's adventures, in their beginning on the last night of 1599. Auriol has been mortally wounded and is brought for treatment to an aged alchemist called Dr Lamb. The character of Lamb is probably based on the notorious John Lambe (1545/6–1628), an astrologer who was arrested for sorcery, Satanism, adultery and child rape, and acquitted, before he was beaten to death by an angry mob on his release.[82] Although the original Dr Lamb was a sixteenth-century figure, he was known in nineteenth-century literature through Sir Walter Scott's description in *Letters on Demonology and Witchcraft* (1830).[83] In Ainsworth's fiction, Lamb is revealed to be Auriol's long-lost grandfather, who has spent his life secretly working to create an elixir of immortality. The old man's appearance is described as that of 'an archfiend presiding over a witches' sabbath', while the firelight in the laboratory 'chamber' nightmarishly transforms its contents, making the gourds 'great bloated toads bursting with venom' and bolt-heads 'monstrous serpents'.[84] The scene of dark magical arts resembles the associations with electricity referred to by other fiction authors, while the phenomenon's hot, popping, distorting nature is prefigured in the sibilance and fitful plosives of the description.

On the night Dr Lamb meets Auriol, he has finally succeeded in creating the long-sought elixir and he is determined to complete and consume it, despite Auriol lying prostrate on the sofa, on the brink of death. Just as Lamb reaches at last for the magic phial, he is struck down – too feeble to reach it. Auriol, desperate to save himself, ignores Lamb's pleas for help, forces him aside and swallows the contents himself, deliberately allowing the old man to die. The illustration from the serialised version of the story (see Figure 1.1) renders the scene in the chamber, where masks and skulls peer out from the walls on the left and a skeleton hovers ominously beyond the curtain on the right.

The moment when Auriol swallows the elixir is portrayed as though he has been electrocuted: 'flashes of light passed before Auriol's eyes and strange noises smote his ears'.[85] The objects in the room 'reeled and danced around him' and the room's phantasmagoric contents appear to come horribly to life, as if to emphasise the elixir's unnatural life-giving qualities.[86]

Darcy, *Les Fontaines Publiques de la Ville de Dijon*, tr. 'The Public Fountains of the Town of Dijon' (Paris: Dalmont, 1856).

[81]     Jenkins (2007), 24.

[82]     Anita McConnell, 'Lambe, John (1545/6–1628)', *Oxford Dictionary of National Biography* (Oxford University Press, 2004) [http://ezproxy.ouls.ox.ac.uk:2117/view/article/15925; accessed 25 October 2010].

[83]     Walter Scott, *Letters on Demonology and Witchcraft: Addressed to J. G. Lockhart* (London: Murray, 1830), 349.

[84]     Ainsworth (1890), 14.

[85]     Ainsworth (1890), 21.

[86]     Ainsworth (1890). The dramatic physical sensations undergone by Auriol may also be an oblique reference to Richard Horne's depiction of the suffering of genius in *The Poor*

Figure 1.1 'The Elixir of a Long Life' (*Auriol*, 1845)[87]

---

*Artist; or, Seven Eye-Sights and One Object* (London: John Van Voorst, 1850), where the protagonist loses his romantic heroine 'Aurelia'.

[87] 'The Elixir of a Long Life' (illustration), *New Monthly Magazine and Humorist*, 74:295 (July 1845), 420.

Auriol is transformed by the elixir into Ainsworth's super-powered hero. In consuming the elixir, he seems not simply *affected* by electricity but also momentarily *inside* it, witness to fundamental changes in apparently solid objects, brought about by force, heat and liquefaction:

> The glass vessels and jars clashed their brittle sides together, yet remained uninjured; the furnace breathed forth flames and mephitic vapours; the spiral worm of the alembic became red hot, and seemed filled with molten lead.[88]

The elixir's fiery, electrifying power transforms Auriol's physical constitution, allowing him to pass through a lofty 'oriel' (or window) in time and occupy apparently contradictory dimensions, protected by the elixir from physical degeneration. Auriol's body is healed and revitalised by the elixir, so that all traces of his earlier wound vanish and he is endowed with 'preternatural strength'.[89] The description demonstrates James Mussell's contention that 'electricity, with its rapid movement through conductors or across space through induction, provides a convenient metaphor for the networks which mobilize and stabilize objects'.[90] The experience also gestures towards the word 'aura', a term used in electrical science to describe the outer envelope of effective influence surrounding various bodies, especially the atmosphere around electrified bodies, and the sphere within which the attractive force of a magnet acts. The word was used in relation to electricity by Benjamin Franklin in the late eighteenth century and by Samuel Taylor Coleridge in referring to an 'electrical aura' in 1810 (OED).

The medical and transformative etymology of the term 'elixir' is also relevant to Auriol's experience, as a treatment created by means of alchemy.[91] Ainsworth's depiction is characteristic of speculations about electrical power throughout the nineteenth century in both non-fiction and fiction. The characterisation of electricity as an elixir and vice versa was not new; in 1839, a *Fraser's* columnist speculates that electricity may actually be the long-sought mysterious power because, not only is it 'as well known to the ancients as ourselves' and the 'hermetic fire of the alchemists' but because medieval researches were also 'close approximations' to modern experiments by Faraday, Fox and Crosse.[92] He proposes, too, that 'this fire was *expressly called electricity*' (author's emphasis).[93] The concept continued to be portrayed in short fiction into the 1870s, for example, when a character is

---

[88]    Ainsworth (1890), 21–2.

[89]    Ainsworth (1890), 22.

[90]    James Mussell, *Science, Time and Space in the Late Nineteenth-Century Periodical Press: Movable Types* (Aldershot: Ashgate, 2007), 189.

[91]    Elixir, *n*: 9a. med. L. *elixir* (cf. Fr. *Elixir*, It. *elissire,* Sp. *elixir*, Pg. *elixir*), ad. Arab. *Al-iks r* (= sense 1), prob. Ad. Late Gr. 'desiccative powder for wounds'] (OED).

[92]    'Alchemy, by An Alchemist', *Fraser's Magazine for Town and Country*, 19:112 (April 1839), 447.

[93]    'Alchemy' (1839).

depicted quenching his thirst and hunger, 'the solids melted in his mouth, the liquid raged through his veins like oil charged with electricity and elixir *vitæ*'.[94] The immortalising force of Ainsworth's fictional elixir can be read as a similar juxtaposition of metaphor, modern technology and medical mysticism.

In *Auriol*, it is 1830 when we meet the hero again, and a beautiful young woman called Ebba Thorneycroft has fallen in love with him. However, Auriol has to give her up to an evil character called Cyprian Rougemont, according to a pact they made in 1800, where they regularly exchange the women who fall in love with Auriol for Rougemont's fabulous riches. After Rougemont takes Ebba away, Auriol goes in search of her, as does her father, accompanied by a gang of street characters who have featured throughout the novel. The chapter in which electricity makes its most significant appearance is called 'The Cell', in what appears to be a reference to an electrical battery. Humphry Davy employed the term 'cell' in electricity from 1801 and, though his usage did not always exactly match the modern term, it did establish as commonplace references to cells in galvanic piles and the technology of later electrical batteries.[95] The décor of the house where Rougemont is hiding hints at his exotic and mystical beliefs, as well as the interests of Ainsworth's readership. The fifth and final cell-like chamber has 'dusky oak' panels, a tapestry 'representing the Assyrian monarch Ninus, and his captive Zoroaster, King of the Bactrians', as well as 'squares and circles' traced on the floor and scattered 'conjuring apparatus'.[96] Rougemont is a Mephistophelean figure whom, we see later, is accompanied by the familiars of a monkey and a dwarf called Flapdragon. The ancient religion of Zoroastrianism was barely known in the West until the end of the eighteenth century.[97] However, from the 1820s, there is evidence of interest from British literary figures such as the poet Felicia Hemans and the Anglican hymnist, Reginald Heber.[98] Ainsworth's depiction may also reflect the increasing awareness of oriental concepts, due to imperial expansion, such as the Hindu idea of *prana* as the underlying vital energy of the universe, which was known in the West from the late eighteenth century.[99]

The dramatic climax of the story draws on some of the most recent scientific developments of the period, even as it is infused with ideas of the archaic. In the chamber, three of Thorneycroft's group sit down in chairs near an electrical

---

[94] 'A Woman-Hater', *Blackwood's Edinburgh Magazine*, 121:738 (April 1877), 415.

[95] June Z. Fullmer, *Young Humphry Davy: The Making of an Experimental Chemist* (Philadelphia: American Philosophical Society, 2000), 312.

[96] Ainsworth (1890), 164.

[97] Mary Boyce, *A History of Zoroastrianism* (Leiden: Brill, 1975), ix.

[98] Mark Knight and Emma Mason, *Nineteenth-Century Religion and Literature: An Introduction* (Oxford: Oxford University Press, 2006), 75. Reginald Heber authored the popular Anglican hymn 'Holy, holy, holy'.

[99] Charles Wilkins, Sir, *Bhagvat-Geeta, or Dialogues of Kreeshna and Arjoon* (London: Nourse, 1785), 113. Wilkins refers to 'the two spirits of Pran and Opan', by which the universe is determined (OED).

machine, the chairs that prove to be the 'enchanted chairs' after which the chapter is named.[100] Suddenly, the men receive a powerful electric shock that forces the weapons from their hands; their arms are pinioned by wooden hooks that jump out of the chair-backs and their legs are caught by fetters that spring out of the floor.[101] Heavy bell-shaped helmets drop from holes in the ceiling, 'fashioned like those worn by divers at the bottom of the sea' with round eyelet-holes of glass and, despite their shouting and swearing, they are all mercilessly 'extinguished'.[102] The type of diving helmet to which Ainsworth refers had only been invented in 1837 and was still in development; the reference gives the scenario a modern twist, as does the technological application of electricity. The electric chair would not be used as a means of execution for another three decades, in the 1880s.[103] Instead, Ainsworth's depiction combines the newest nineteenth-century diving technology with his own imaginings of electricity's potential. Yet his interest is not focused entirely on the future and technology. The adjective 'enchanted' endows the chairs with archaic, magical powers and they are 'large, high-backed and grotesquely-carved', like instruments of medieval torture.[104] The fashionable Gothic design of the electrical chairs makes them a meeting-point between the story's different time-frames of the sixteenth and nineteenth centuries. However, it is Ainsworth's imagined application of their use as lethal electrical instruments, which makes them truly macabre.

In the story's rather unsatisfactory conclusion, Rougemont reveals that Ebba has never existed and he supplies an antidote elixir. To Auriol's astonishment, he finds himself back in the original chamber, which is 'precisely as he had seen it above two centuries ago' and where his grandfather is alive and well, working on the elixir with no memory of what had happened with Auriol. Auriol concludes that his experiences were nothing but the delirium of a fever, but also senses that he has lived for centuries in a few nights. It is an uninventive and somewhat cursory ending, although one fittingly suggestive of medieval dream visions, such as *Piers Plowman* or *Pearl*.

While *Auriol* can be shown to engage with particular aspects of electricity and its applications, contemporary scientific practice is still predominantly on the margins of the novel. The role of electricity in the novel is emblematic, placing it closer to the metaphorical responses than to the progress of actual mid-century scientific investigations. The technique uses fiction to distance the reader from the realities of science and experiment. Dr Lamb's egocentric and medieval alchemy appears to contradict the altruistic pragmatism of such practitioners as Anthony Peck and William Sturgeon and the atomic investigations of Faraday and Maxwell.

---

[100]    Ainsworth (1890), 163–4.

[101]    In fact, powerful electric shocks cause muscle contraction, rather than release.

[102]    Ainsworth (1890), 176, 177.

[103]    Mark Regan Essig, *Edison and the Electric Chair: A Story of Light and Death* (Stroud: Sutton, 2003).

[104]    Ainsworth (1890), 163.

*Auriol* appears designed to appeal to contemporary nostalgia for the inexplicable; yet, like the romantic history of 'Dr. Beroni's Secret', electricity is also portrayed as a power that offered an innate possibility of escaping limitations of location, time and individual circumstance.

Until the 1870s, very few novels referred at length or directly to electricity; it tends to be mentioned in the form of metaphorical or implied references, with few instances of extended or meaningful engagement with the subject. We might ask, therefore, why there are not more sustained engagements with electricity, particularly after several decades in which the subject had been studied so systematically. Why were the subsequent years not bristling with electrical stories? Acknowledging the rarity of novels specifically about electricity in the period is an important part of understanding literary responses to contemporary scientific developments. The historical progression of electrical science is key – showing, again, the extent to which literary analysis can be vitally informed by the history of science. As discussed earlier, before the 1870s, electricity was barely understood sufficiently by specialist scientists for it to be represented, let alone responded to coherently, and fiction authors were generally at greater distance from explorations and discoveries in the field than scientists. In the 1870s and 1880s, fiction authors appear to have become sufficiently comfortable with the subject to incorporate it into recognisable literary genres and, at that point, the distinction between writing about electricity and science fiction breaks down almost entirely. As illustrated by the novels discussed in the rest of this chapter, electricity became then more commercially viable as a source of popular literary sensation.

The persistent association between elixirs and electricity is especially apparent in Edward Bulwer-Lytton's novel *A Strange Story* (1862).[105] The characterisation of the 'elixir of life' draws on the author's combined interest in the occult, mesmerism and electricity.[106] Initially, the novel's narrator, Allen Fenwick, is a determinedly materialist doctor with a recreational interest in electricity; however, from the outset, the workshop in which he keeps the electrical battery for his hobby is also described as his 'sanctuary'; 'the house-maid was forbidden to enter it with broom or duster except upon special invitation' and it is 'cut off from the main house', establishing electricity and experimentation as liminal pursuits that are entirely separate from home, family life and professional occupations.[107] Fenwick is helped in operating the new battery by the peculiar yet charismatic Margrave, who proposes that the human body is an integral part of the electrical circuitry and attributes the failure of the battery to a burn on Fenwick's hand, which interrupts

---

105 Edward Bulwer-Lytton, *A Strange Story* (Boston: Gardner A. Fuller, 1862). The novel was serialised in the weekly *All the Year Round* the year after Dickens's *Great Expectations*, under the pseudonym 'The Author of My Novel Rienzi'. In 1861, *All the Year Round* was priced at 2d and its circulation rose from 120,000 (1859–1860) to 300,000 (1869) per issue.

106 Bulwer-Lytton (1862), 376.

107 Bulwer-Lytton (1862), 85.

the circuit because 'the least scratch in the skin of the hand produces chemical action on the electric current'.[108] Fenwick is impressed that 'in a moment the needle of the galvanometer responded to [Margrave's] grasp on the cylinder', but he also laughs at how Margrave's knowledge of electricity relies arbitrarily on both new and archaic sciences – a narrative voice that indicates Bulwer-Lytton's awareness of his readers' different levels of knowledge.[109] The quest for an immortalising elixir that dominates the novel, together with the inclusion of mysterious, veiled and oriental characters, such as Ayesha and the Syrian Haroun, reflects Bulwer-Lytton's fascination with nineteenth-century interpretations of Eastern mysticism, as well as the awkward boundaries of emerging Western sciences.[110]

As the novel progresses, it is revealed that Margrave has gained immortality from ingesting the secret, liquid contents of a 'glittering phial', an intriguing transformation that makes him, like the other electrically animated creatures of the period's fiction, a fusion of the human and disturbingly non-human.[111] At the museum with Margrave, Fenwick is cast into a spell-like trance by Sir Philip Derval, which lets him see inside Margrave's body. In the trance, he is transported back in time and can see the red and azure lights of Margrave's original constitution, as well as the bright, indestructible 'silvery spark' that, he notes, was absent in the museum's animal specimens when they were alive.[112] Just as he realises that the silver light is a halo around the soul, rather than the soul itself, the light disappears, leading Fenwick to conclude that Margrave's soul has vanished in the transformation to immortality and that, as a result, his being has become ruled by grossly distorted moral, mental and sensual faculties. Bulwer-Lytton's portrayal of the association between electricity, elixirs and experimentation is more elaborate and complex than its presentation in the short fictions or *Auriol*; it affirms equally, however, the disfiguring repercussions of their juxtaposition for man's moral character and for the natural, ordered balance of the physical.

The life-giving elixir is described repeatedly in terms of light, heat and electricity, particularly in the climactic ceremony to re-create it in a cauldron, significantly surrounded by lights meant to keep at bay other dark and supernatural forces. The ritualistic and occult scene resembles the séance in the short story 'Doctor Beroni's Secret', while the 'dazzle' and 'flash' of the 'molten red' potion is reminiscent of its representation in Ainsworth's novel.[113] During the rite, Margrave's crazed behaviour resembles particularly that of Elias Johns or Melchior, in the short stories discussed in Chapter 4, as he murmurs gleefully, 'see the bubbles of light, how they sparkle and dance! I shall live! I shall live!' and, to

---

[108]    Bulwer-Lytton (1862), 87.
[109]    Bulwer-Lytton (1862).
[110]    Bulwer-Lytton's interest in Roscrucianism, for example, is made explicit in the title and focus of his previous novel, *Zanoni: A Rosicrucian Tale* (1842).
[111]    Bulwer-Lytton (1862), 166.
[112]    Bulwer-Lytton (1862), 129.
[113]    Bulwer-Lytton (1862), 377.

his ultimate cost, he forces his way forward to secure the elixir for himself.[114] His greedy desperation mirrors Auriol's, and reaffirms the psychological imbalance and moral bankruptcy portrayed as inseparable from the downfall of those who become too closely involved with experimentation, elixirs and electricity.

The distinctive feature of *A Strange Story*, as a literary response to electricity, is its additional framework of contemporary explorations of electricity and matter, which accompanies the depiction of the immortalising elixir. Scientific figures and works are referenced throughout the narrative and in footnotes, and include the electro-physiologist Emil du Bois-Reymond, Davy's 'Heat, Light and the Combinations of Light', Hamilton's 'Lectures on Metaphysics and Logic', and the research of Alexander Bain and Hare Townsend.[115] The references may have contributed a degree of credibility to the otherwise improbable tale; certainly, they introduce an extended non-fiction element to the fiction and connect it directly to the more distinctly scientific investigations of the period.

The inspiration for Bulwer-Lytton's depiction of Margrave may also have included the work of James Hinton, a leading aural surgeon and philosopher, whose series of studies on 'physiological riddles' was published in several issues of the *Cornhill Magazine* just a year before the publication of *A Strange Story*.[116] Hinton considered the origins of what he referred to as the 'active power' of the animal body and envisaged the body itself as a machine.[117] Hinton proposed also that the body's equilibrium is constantly overthrown by various stimuli and that its functions are, therefore, a 'ceaseless round of force-mutation throughout nature, each one generating, or changing into, the other'.[118] He also alleged that the laws of energy conservation provided 'the plan on which the animal creation is constructed'.[119] Hinton's depiction of the body's power and forces as a form of electrical circuitry provides important scientific context and validation for Bulwer-Lytton's characterisation of Margrave as electrically charged. While Bulwer-Lytton does not appear to have had any official affiliation with the *Cornhill*, it was one of the century's best-selling periodicals, with a circulation that reached 87,500 a month by 1860.[120] *A Strange Story* is, Andrew Brown suggests, characteristically 'underpinned' by Bulwer-Lytton's 'voracious reading in the relevant scientific and

---

[114]  Bulwer-Lytton (1862), 114.

[115]  Bulwer-Lytton (1862), 114, 294, 295.

[116]  James Hinton (1822–1875), 'Physiological Riddles. I. – How We Act', *Cornhill Magazine*, 2:1 (July 1860), 21–32; 'Physiological Riddles. II. – Why We Grow', 2:2 (August 1860), 167–74; 'Physiological Riddles. III. – Living Forms', 2:3 (September 1860), 313–25; 'Physiological Riddles. IV. – Conclusion', 2:4 (October 1860), 421–31. Science in the Nineteenth-Century Periodical: An Electronic Index, v. 3.0 [hriOnline http://www.sciper.org; accessed 3 May 2011].

[117]  Hinton (July 1860), 21.

[118]  Hinton (July 1860), 23.

[119]  Hinton (July 1860), 24.

[120]  Gowan Dawson, 'The Cornhill Magazine and Shilling Monthlies in Mid-Victorian Britain', in *Science in the Nineteenth-Century Periodical: Reading the Magazine of Nature*,

philosophical literature'.[121] The combination of the two factors makes it seem at least likely that Bulwer-Lytton would have read or heard about Hinton's depictions.

Responses to electricity were also responses to technological progress and to the century's new understandings of functions and systems; as Louise Henson suggests, in the 1860s period of advancement in telegraphy and early photography, 'metaphors associating the mediatory physiological processes of the body with those of technological activity abounded'.[122] Bulwer-Lytton juxtaposes new and old concepts, explanations and metaphors. As he observed, 'art and science have their meeting point in method'; it is only that 'in art, method is less perceptible than in science'.[123] In fiction, the 'method' of electrical science was not essential for a story to be understood and Bulwer-Lytton's purpose was not necessarily one of interpretation or explanation; nevertheless, the ideas of science and speculation about its procedures influence and often act as a crucial impetus for his fiction.

## Imagining Electrical Power

In *The Coming Race* (1871), Bulwer-Lytton interrogates the possibilities of electricity further and steps away from the medievalist supernaturalism of Ainsworth's response, as well as that of his own earlier novels. In *The Coming Race*, electricity plays a significant role and one that interacted with the whole series of new technologies beginning to emerge in the second half of the century, such as the electric telegraph (1858), the telephone (1876), the light bulb (1879), the electric train (1879), the car (1885) and the radio (1895). The innovations depicted in the novel indicate how substantially electricity had, by the beginning of the 1870s, already transformed perceptions of man's potential for communication and movement. As Herbert Sussman suggests in his seminal discussion of the impact of Victorian technology, in the nineteenth-century technological revolution, animal and human 'muscle power' was not just transformed but also replaced by the energy generated by steam and electricity.[124] The changes happened at a pace never experienced before and the speed technology could provide was increasingly

---

ed. Geoffrey Cantor, Gowan Dawson, Graeme Gooday, Richard Noakes, Sally Shuttleworth and Jonathan R. Topham (Cambridge: Cambridge University Press, 2004), 126.

[121]    Andrew Brown, 'Lytton, Edward George Earle Lytton Bulwer, First Baron Lytton (1803–1873)', *Oxford Dictionary of National Biography* (Oxford University Press, 2004) [http://ezproxy.ouls.ox.ac.uk:2117/view/article/17314; accessed 28 August 2011].

[122]    Louise Henson, '"In the Natural Course of Physical Things": Ghosts and Science in Charles Dickens's *All the Year Round*', in *Culture and Science in the Nineteenth-Century Media*, ed. Louise Henson, Geoffrey Cantor, Gowan Dawson, Richard Noakes, Sally Shuttleworth and Jonathan R. Topham (Aldershot: Ashgate, 2004), 117.

[123]    Edward Bulwer-Lytton, Baron, 'On Certain Principles in Art in Works of Imagination', *Miscellaneous Prose Works*, vol. 3 (London: R. Bentley, 1868), 348.

[124]    Herbert Sussman, *Victorian Technology: Invention, Innovation, and the Rise of the Machine* (Santa Barbara: ABC-CLIO, 2009), 5.

sensationalised. *The Coming Race* was unashamedly sensationalist but, like some of the fictions already discussed, the novel also envisages the repercussions of electricity, technology and their effects upon society and human nature.

In *The Coming Race*, the unnamed narrator and his friend, an engineer, attempt to investigate some lights beneath a fresh mine shaft, and discover a vast system of underground caverns. The novel's publication in the same year as Charles Darwin's *Descent of Man* (1871) makes it possible that the title and opening depiction are thinly-veiled references to evolutionary competition between and within species and, as a social satire on contemporary society, the novel certainly engages with these issues and others.[125] The most recent edition of the novel emphasises the novel's connection with the 1870s revival of the 'hollow earth' discourse and Peter Sinnema describes it as 'the crowning achievement' of the hollow earth genre.[126] The theory was initiated by the seventeenth-century astronomer Edmond Halley and revived in 1871 through the publication of a 'treatise' by an American military man, John Cleves Symmes, Jr. (1780–1829), who proposed that the poles were the earth's openings and lobbied for funding to lead an expedition to the planet's interior throughout the 1820s.[127] Sinnema also suggests that the theory had 'become something of a vogue in certain learned and literary circles' by the 1820s, indicating a level of dilettantism that means its relevance should not be overstated.[128] *The Coming Race* was published in 1871, however, half a century after the idea had peaked in British popular literary culture and by which time interest was much more marginal, as indicated by the theory's 'religio-philosophical' publication. Like Bulwer-Lytton's unfinished posthumous novel, *Parisians* (1873), *The Coming Race* is predominantly a satire upon 'frivolous society', a key feature of which was the rapidly increasing technological advancement of electricity.[129]

As the two men make their way down the mine shaft, the engineer falls to his death with the climbing equipment, leaving the narrator alone with no means of escape. He soon has his first encounter with one of the occupants of the underworld, the *Vril-ya*, who are highly intelligent and live in a brightly lit, cavernous underworld. The source of power is 'vril' and it is not precisely

---

[125] See Patrick Brantlinger, 'Race and the Victorian Novel', in *The Cambridge Companion to the Victorian Novel*, ed. Deirdre David (Cambridge: Cambridge University Press, 2001), 149–68.

[126] Edward Bulwer-Lytton, *The Coming Race*, ed. Peter W. Sinnema (1871; repr. Peterborough: Broadview Press, 2008), 9.

[127] Edmond Halley, 'An Account of the Cause of the Change of the Variation of the Magnetic Needle; with an Hypotheses of the Structure of the Internal Parts of the Earth', *Philosophical Transactions of Royal Society of London*, 16 (1692), 563–78; M.L. Sherman and William F. Lyon, *The Hollow Globe: The World's Agitator and Reconciler a Treatise on the Physical Conformation of the Earth* (Chicago: Religio-Philosophical Publishing House, 1871). See also, P. Clark, 'The Symmes Theory of the Earth', *Atlantic Monthly* 31 (April 1873), 471–80.

[128] Bulwer-Lytton (2008), 9.

[129] 'The Parisians', *Edinburgh Review*, 139:284 (April 1874), 383.

synonymous with electricity; indeed, as Eckhart Voigts-Virchow argues, 'it is not a metaphor for electricity, but for the potential of electricity'.[130] Electrical power in the form of 'vril' has also distorted the *Vril-ya*'s physical appearance, although to a lesser extent than the gnarled and twisted tree discussed in 'The Tree of Knowledge' (1853), an anonymous short story discussed in Chapter 4. The *Vril-ya*'s ancestors were frogs, in what is possibly an oblique reference to Galvani's original experiments, but they have become extremely tall with 'large black eyes, deep and brilliant'; they have the shared expression of a 'sculptured sphinx' and each one has wings 'folded over its breast and reaching to its knees'.[131] Although they stand upright like humans, with the addition of their electrical wings, they also resemble bats, creatures usually allied to darkness. The initial hybridity of their genetic origins has progressed through their development of a power similar to electricity, which allows them to fly. While they are not portrayed as repellent, the combination represents a distortion of otherwise logical, 'natural' or even evolutionary developments.

The *Vril-ya*'s wings, we learn later, are 'electric contrivances', which not only give them the ability to fly but also demonstrate their mastery of electrical technologies, reflecting contemporary interest in electricity as a means to develop controlled and manned, sustainable flight. When the narrator compares the *Vril-ya* to birds, it is the expertise of their flight that he admires, noting that it is 'as swift as an eagle's' and that they perform complex flying modes with ease, swooping, hovering and 'undulating' with 'fantastic grace'.[132] Long before the nineteenth century, manned flight had been contemplated in ancient Greece, China, and Leonardo da Vinci's fifteenth-century sketches of 'ornithopters'.[133] However, from the 1830s, electrical magnetos provided the promise of reliable ignition systems and real progression in aeronautical engineering, although the four-stroke internal combustion engine would not be invented by N.A. Otto until 1876.[134] Electrical technologies could transform flight from the stuff of wishful thinking and fantasy to scientific reality. The British Aeronautical Society of Great Britain was founded in 1866 and constructed the first research wind tunnel in 1871, the same year as Bulwer-Lytton's novel, as a means to derive 'data on which a true science

---

[130]    Eckhart Voigts-Virchow, 'Melancholy Elephants and Virgin Machines: Technological Imagery and Mechanical Lacunae from Industrial Novels to Scientific Romances', in *Lost Worlds and Mad Elephants: Literature, Science and Technology, 1700–1990*, ed. Elmar Schenkel and Stefan Welz (Leipzig: Galda and Wilch, 1999), 154.

[131]    Edward Bulwer-Lytton, *The Coming Race*, ed. Matthew Sweet (London: Hesperus Press, 2007), 9.

[132]    Bulwer-Lytton (2007), 13, 16.

[133]    See Richard P. Hallion, *Taking Flight: Inventing the Aerial Age from Antiquity through the First World War* (Oxford: Oxford University Press, 2003), 25.

[134]    Laurence Goldstein, *The Flying Machine and Modern Literature* (Bloomington: Indiana University Press, 1986), 54.

of aeronautics can be founded'.[135] At the same time, the French led the way as aviation pioneers during the 1870s, and Victor Hugo proposed the flying machine or '*aéroscaphe*' as a crucial part of a 'future divine and pure'.[136] The concurrent development of sciences relating to the technological development of electricity enhanced interest in the subject further and made the possibilities depicted in fiction seem all the more feasible.

As we see in relation to non-fiction periodical writings, manipulating electricity did not necessarily mean understanding it. In *The Coming Race*, the opposite is the case: the *Vril-ya* mastery of 'vril' is inseparable from its influence within their highly developed culture. They are portrayed as entirely the opposite of the volatile characters often connected with electricity in short fiction. The *Vril-ya*'s countenances are described as 'smoothly serene', their demeanour 'grave and courteous', and their interactions as 'stately'; indeed, they 'despise any vehement emotional demonstration'.[137] Like Rougemont's house in *Auriol*, the influence of colonialism is apparent in the 'oriental splendour' of the *Vril-ya*'s homes, as well as the reference to their having the 'gravity and quietude of the Oriental'.[138] It is precisely this detached composure that makes them so adept in mastering the phenomenon.

The contrast between the two types of behaviour is especially apparent shortly after the narrator receives an electric shock from touching his host's wing. The narrator's behaviour quickly deteriorates into the type of mental and physical imbalance represented in short fiction portrayals of electrical practitioners. First, he feels his mind begin 'to wander' and then suffers a sudden onset of 'terror' and 'wild excitement', repelling his kindly host's concern 'with vehement gesticulation, and forms of exorcism, and loud incoherent words'. Finally, despite being shown that the wings are 'but a mechanical contrivance', he springs at his host's throat 'like a wild beast', and has to be 'felled to the ground' by a further electric shock.[139] The narrator's regression in response to electricity represents an early example of later Victorian degenerationist fears, where man's uncontrollable savagery was perceived to lie just beneath his apparently civilised appearance, waiting to emerge, perhaps as the direct result of 'over-civilisation'.[140] In *The Coming Race*, as Barri Gold points out, electricity has 'brought about peace

[135]  Nils Henrik Randers-Pehrson, *History of Aviation* (New York: National Aeronautics Council (US) Inc., 1944), quoted in Hallion (2003), 116.

[136]  Victor Hugo, *Légende de Siècles*, tr. 'The Legend of the Centuries' (1859; 1877; 1883), quoted by Goldstein (1986), 5. Key aviation pioneers included Jean-Marie Le Bris (1817–1872) and Alphonse Pénaud (1850–1880).

[137]  Bulwer-Lytton (2007), 12, 13, 15.

[138]  Bulwer-Lytton (2007), 11.

[139]  Bulwer-Lytton (2007), 17.

[140]  Deirdre David, 'Sensation and the Fantastic in the Victorian Novel', in *The Cambridge Companion to the Victorian Novel*, ed. Deirdre David (Cambridge: Cambridge University Press, 2001), 209.

through the capacity for total annihilation'.[141] In that sense, electricity retains the dual character that authors repeatedly emphasise; although it is a unifying force, it is simultaneously a means of civilisation and degeneration, healing and harm, creation and destruction.

The *Vril-ya*'s world epitomises its inhabitants' serenity, grace and efficiency but it also has more eerie qualities. The mechanical automatons that perform menial services stand motionless yet poised at various points along the walls; they are animated by electricity and it gives them the appearance of 'living form', in an updated version of electricity as a life-giving elixir. Despite the narrator knowing they are machines, he admits that he cannot help thinking of them as 'dumb and motionless' and 'phantom-like'.[142] They have agency but they are not personified; in fact, for him, they are confusingly non-human. Like the 'unseen stream' experienced by Lucy Snowe in *Villette*, they have 'a rapid and gliding movement, skimming noiselessly over the floor'.[143] Just as electricity transforms the *Vril-ya*'s potential for movement, it endows machines with an artificial, preternatural speed and smoothness, serving to emphasise the narrator's own organic and 'natural' basis.

Electricity accentuates the stark contrasts in the novel between man and machine, but there is also a notable *lack* of contrast in the underworld's electric light. The narrator remarks that there is a 'serene lustre diffused over all by a myriad of lamps', making the underworld seem both 'splendid' and sombre, 'lovely' as well as 'awful', and lending the vista a 'wild and solemn beauty impossible to describe'.[144] His contrasting impressions convey the contradictions that responses to electricity so often involved. The application of electricity to lighting took place within a context of increasing visual awareness.[145] The light it gave was appealing but, in its artificiality, it was also persistently disturbing. Bulwer-Lytton's portrayal of electric light as a measure of species development offers further moral associations, within the context of the novel's social satire. The *Vril-ya* have a distaste for darkness and associate it with ignorance, primitivism and disgust; they refer to the 'primeval savages' and 'barbarous tribes' who live in the most 'desolate and remote recesses of uncultivated nature', unacquainted with other light than that they obtain from volcanic fires, and 'contented to grope their way in the dark, as do many creeping, crawling and flying things'.[146] As Edward

---

[141]    Barri J. Gold, *ThermoPoetics: Energy in Victorian Literature and Science* (Cambridge, MA: MIT Press, 2010), 82.

[142]    Bulwer-Lytton (2007), 11, 15.

[143]    Bulwer-Lytton (2007), 10. The electric 'stream' relies again on the fluid metaphor, as does George Eliot's reference, when she writes that 'an electric stream went through Dorothea', *Middlemarch*, ed. Rosemary Ashton (London: Penguin, 2003), 37.

[144]    Bulwer-Lytton (2007), 16.

[145]    See Kate Flint, *Victorians and the Visual Imagination* (Cambridge: Cambridge University Press, 2000).

[146]    Bulwer-Lytton (2007), 19.

Beasley notes, in the predominantly black and white illustrations appearing in the nineteenth-century press from around the world, particularly the colonies, people were 'graded along a single continuum of lightness and darkness'.[147] Light and dark increasingly signified civilisation and savagery, and not just in respect of colonial races. In Charles Booth's mapping of London's poverty in the 1880s, for example, rich and poor areas were denoted by lighter or darker shading. Like these, Bulwer-Lytton's novel contributed to understandings of light as a feature related to civilisation, just as the *Vril-ya* suggest. To this, movement was again added as a means by which different forms of life were distinguished from one another, in terms of their level of civilisation and material nature. From the racehorse's gait to lady-like deportment, smooth movement suggested superior cultivation and advancement. Electricity provided a way of achieving smooth, silent movement; however, its artificiality also tapped into age-old fears, of things that move without human agency. The new dimensions made possible by electricity problematised the 'natural' developmental features of civilisation, against which Bulwer-Lytton's narrator constantly assesses the rank of his own species.

Although the power wielded by the *Vril-ya* is called 'vril' and the narrator suggests that it is almost synonymous with electricity, it also includes numerous 'other forces of nature', providing a form of 'unity in natural energetic agencies' referred to as 'magnetism, galvanism, etc.'.[148] By the early 1860s, galvanism was already widely recognised as 'long ago obsolete', albeit valuable in terms of theory and history.[149] The narrator adds the more up-to-date reference of Michael Faraday's 'forces of matter', which have 'one common origin' and are 'convertible into one another'.[150] The reference to common origins occurs immediately after the narrative focus on different life forms and species, aligning issues of evolution and biology with those of physical matter and electrical theory. The reference to electricity's scientific development in the midst of a work of fiction bears testament to the afterlife and literary discourse around Faraday, as well as the impact of his research. Although Faraday died in 1867, several years before the publication of *The Coming Race*, it is he whom Bulwer-Lytton quotes rather than any of the other scientists who were still active in the field.[151] Figures such as John Tyndall, for example, were conducting tremendously popular lectures on electricity at the Royal Institution, which were available in published form while

---

[147]    Edward Beasley, *The Victorian Reinvention of Race: New Racisms and the Problem of Grouping in the Human Sciences* (London: Routledge, 2010), 15.

[148]    Bulwer-Lytton (2007), 22.

[149]    Joseph Frick, *Physical Technics: or, Practical Instructions for Making Experiments in Physics and the Construction of Physical Apparatus with the Most Limited Means* (Philadelphia: J. B. Lippincott and Co., 1861), 315.

[150]    Frick (1861).

[151]    Other leading electrical scientists who were especially active at the time included James Joule, Balfour Stewart, James Clerk Maxwell, George Gabriel Stokes and William Thomson.

Bulwer-Lytton was writing *The Coming Race*.[152] Indeed, Bulwer-Lytton quotes the same passage that was re-produced in Tyndall's *Faraday as a Discoverer* (1868) so it is possible that Bulwer-Lytton even referred to Tyndall's volume.[153] At the same time, Faraday's celebrity meant that there was no need to explain to readers who he was, and, of course, as he was no longer living, he could not object to the associations Bulwer-Lytton attached to him.

The narrator emphasises that 'vril' is used 'scientifically' and that it gives the *Vril-ya* power over the weather, minds and bodies 'to an extent not surpassed by the romances of our mystics'.[154] Throughout the novel, the narrator is at pains to present himself as rational, systematic and sceptical, in a way that indicates Bulwer-Lytton's wish to distance himself from fashionable pseudo-sciences. The narrator's internal monologue, for example, is studiously scientific in tone; he refers to his 'cerebral organisation' rather than his mind or brain, and he tells the *Vril-ya* girl, Zee, that the effects of electricity 'upon certain abnormal constitutions' had been 'fairly examined and analysed' in his own world and that the effects were found to be 'very unsatisfactory'.[155] Zee replies that when her own race was in 'the infancy of their knowledge' about 'vril', there were 'similar instances of abuse and credulity' about the possibilities it offered for flight, telepathy and healing, as well as social reformation.[156] The implication is that the underworld's development has intellectually surpassed the narrator's world and that, in the actual Victorian world, electricity could be developed in the same way, for good or ill. However, the *Vril-ya*'s knowledge of 'vril' gives them telepathic and mesmeric powers, medical practices that were as experimental as electrical treatments in the 1870s. As Fred Kaplan points out, 'mesmerism particularly fascinated an age almost obsessed by the possibility of curing all illnesses and that suffered various epidemics of its own, particularly plagues of the nervous system and the psyche'.[157] Like these other practices, in Bulwer-Lytton's underworld utopia, electricity is represented as Janus-like in character: it could be the cause of nervous conditions and distortions but also their possible cure.

*The Coming Race* ran through eight editions in 18 months and was reissued 25 times, in multi-volume collections of Bulwer-Lytton's complete novels in Britain and America between 1875 and 1900.[158] While publication figures of this scale

---

[152]    For example, John Tyndall, *Light and Electricity* (London: D. Appleton and Company, 1871) and *Notes of a Course of Nine Lectures on Light* (London: Longmans, Green, 1870).

[153]    John Tyndall, *Faraday as a Discoverer* (London: Longmans, Green and Co., 1868), 81.

[154]    Bulwer-Lytton (2007), 22, 23.

[155]    Bulwer-Lytton (2007), 23.

[156]    Bulwer-Lytton (2007).

[157]    Fred Kaplan, 'The Mesmeric Mania: The Early Victorians and Animal Magnetism', *Journal of the History of Ideas*, 35:4 (October–December 1974), 693.

[158]    Brown (2004).

indicate the novel's popularity, they do little to confirm whether readers' interest was prompted by Bulwer-Lytton's response to electricity. The reception history provides more detailed insights into the basis of the novel's popularity and allows us to gauge how far its literary success was related to the portrayal of electricity. The novel appears to have been read primarily as a satire on contemporary society but, as an early reviewer remarks,

> Readers who could hardly have borne to be told in plain language that they were a loathsome and contemptible race of beings could sympathize with the scorn directed against their grotesque representatives.[159]

The novel was published anonymously but its misanthropy was considered to be relatively palatable, in the tradition of Jonathan Swift, with whose work the novel was regularly compared.[160] Without doubt, it is the social, moral and behavioural improvements in *Vril-ya* society that are most frequently asserted to be the novel's most interesting features. The technological advances portrayed in the novel were sometimes treated with irony, as stylised features of the utopian genre; the same *Saturday Review* writer sardonically observes, for example, that the *Vril-ya* 'can of course fly – a capacity which is in great request amongst inhabitants of ideal worlds'.[161] The causal connection between the *Vril-ya*'s mastery of electricity and their social sophistication was also quickly identified, in that their 'progress in the line of physical discovery, has profoundly modified their relations as individuals and members of society'.[162] The development of technological prowess was increasingly understood to affect not just the individual but also the way in which society fitted together.

Bulwer-Lytton's apparent promotion of electrical advancements repeatedly demonstrates his discomfort, particularly as a feature of social progress. This was apparent to readers too; as another reviewer asserts with unmistakable irony, 'eradicate all human instincts; nip all human aspirations in the bud; remove all that makes life worth having; and then, but not sooner, expect to realise your democratic ideal'.[163] It is this feature of the novel that modern scholars describe as a 'Romantic streak of ambivalent technophobia', identifiable in other novels published at much the same time, such as George Chesney's *The Battle of Dorking: Reminiscences of a Volunteer* (1871) and Samuel Butler's *Erewhon* (1872).[164] Imitations of *The Coming Race*'s dystopianism were recognised quickly thereafter. When the novel *Colymbia* (1873) was set in the South Sea islands, for example, it was remarked that 'soon every part of the world, or perhaps of the solar system, will have its

---

[159]   'The Coming Race', *Saturday Review*, 31:813 (27 May 1871), 674.
[160]   'The Coming Race' (1871).
[161]   'The Coming Race' (1871).
[162]   'The Coming Race', *Athenaeum*, 2274 (27 May 1871), 649.
[163]   T.H.S. Escott, 'Bulwer's Last Three Books', *Fraser's*, 9:54 (June 1874), 765.
[164]   Voigts-Virchow (1999), 145.

mysterious colony of grotesque inhabitants'.[165] Electricity was recognised by the novel's readership as central to the ostensibly advanced *Vril-ya* society – not just a result of advancement but also an important determining factor of it.

The effect of electricity on the *Vril-ya*'s individual and collective nature is invisible but pervasive, like the nature of the phenomenon. The social vision portrayed by the novel was as exotic as the applications of electricity, most of which had yet to be thought of, let alone heard of. The combination of electricity's possibilities and the social advancements portrayed in *The Coming Race* set a precedent for depicting electricity as a technology of the future, one that might alter human capabilities and, literally, man's place in the world. Investigations of quality artificial lighting, reliable flying machines and faster communications were the focus of technological innovation and the stuff of fiction; they took place not simply on earth but also deeply within it. As such, like the development of electricity itself, they represented what lay beneath the earth's visible surface and revealed the very centre of human existence.

The comparison of references to electricity in popular nineteenth-century novels results in important implications for our understandings of how literary status is acquired, how works become classified as 'science fiction' and, more widely, how their reception is influenced by scientific associations. The canonical works discussed earlier in the chapter engage with ideas of electricity by means of metaphor, in ways that illustrate how leading novelists responded to the phenomenon. They might appear to offer little in the way of reflection about its technical properties, its contemporary significance, or its role within the broader narrative of each novel. What they do reveal, however, is electricity's role as a signifier of volatility, speed and invisibility and, more unexpectedly, how novelists exploited this to convey the immediacy of emotions such as love and fear, as well as metaphysical connections. The works discussed have tended to retain their literary credibility and worth, despite and even because of the oblique and figurative ways they referred to electrical concepts. They show how electricity's properties could be used to represent possibilities and repercussions both connected to the phenomenon and ostensibly beyond it. Many resemble the short stories about electricity also discussed, not simply in scope or ephemerality, but rather in their lack of enduring status as literary works.

Engaging with electricity appears to have had a direct effect upon the categorisation and reception of the novels discussed. As we know from both non-fiction and fiction writings, electricity was often closely allied to nineteenth-century science, industry and trade. These non-literary and commercial connotations seem also to have brought about the reduced literary status of novels that foregrounded electricity or wrote about it more directly. Indeed, scientific associations so impinged on how the novel was perceived that they were often quickly pigeonholed – or even relegated – to the lowlier genre of 'science fiction'. The processes and their impact have significant consequences for how

---

[165]    'Colymbia', *Saturday Review*, 35:905 (1 March 1873), 289.

we understand science to have influenced literary fiction. At stake is whether the scientific connections of subjects such as electricity, which are often as perceived as marginal to the literary nature of fiction, might in fact have played a much greater role in the literary ranking, reception and longevity of individual works. In order to explore these issues effectively, we need to appreciate how awareness and understandings of electricity developed in the nineteenth century in scientific circles and beyond, aspects addressed in the next chapter.

# Chapter 2
# Electrical Analogies, Science and Poetry

The 'Torch of Science', writes Thomas Carlyle, has been 'brandished and borne about, with more or less effect, for five thousand years and upwards' and now burns 'perhaps more fiercely than ever'.[1] The Olympian metaphor portrays a smooth progression of scientific interest, which culminates in the passionate engagement of his own era. In reality, scientific practice underwent several unique and profound conceptual transitions in the nineteenth century, in which the investigation of electricity was central, and which can be understood as a narrative palimpsest of discovery, experiment and representation.

In considering nineteenth-century literary responses to electricity, it is important to take into account what information was available about electricity, both before and after the 1830s when scientific study of the phenomenon began in earnest, how that information developed and what questions it presented for literary representation. This groundwork also establishes how far writers chose to observe and incorporate the factual information available at the time. The electrical pioneers, Michael Faraday (1791–1867) and James Clerk Maxwell (1831–1879), were all too aware of the overlaps, tensions and harmonies that existed between science, writing and representation, and the ways in which they could use them to conceptualise the phenomenon.

## Electricity's Sensational and Imponderable History

Referring closely to the key scientific developments of the nineteenth century provides us with essential clues, to clarify what might otherwise remain puzzling disparities, proximities and affinities between the period's scientific and literary developments. Approaching discoveries about electricity in this way allows us to see, too, that they did not remain within a firmly bounded scientific world, and contributed, instead, to a dynamic and highly creative relationship between scientists, non-specialists and fiction authors.

Throughout the nineteenth century, and particularly its early stages, there were many competing theories about the nature of electricity, which have since provided

[1]    Thomas Carlyle, *Sartor Resartus: The Life and Opinions of Herr Teufelsdröckh in Three Books*, ed. Rodger L. Tarr and Mark Engel (1831; repr. Berkeley: University of California Press, 2000), 3.

the focus for important and relevant studies.[2] In one example, Geoffrey Cantor highlights the rhetorical qualities of scientific narratives and reports – writings that actively sought to influence and persuade readers – and, as he points out, 'unless we are to believe that truth is manifest we need to view rhetoric as an integral part of science'.[3] The extent to which a scientific concept is considered to be 'true' is so heavily contingent upon its representation within a specific historical context that it would be deeply misguided to separate the form of representation from the concepts it attempts to represent. As David Gooding notes, representations 'can be articulated as *instrumentally* useful concepts before they are incorporated into a theoretical framework, so it is plausible to suppose that they shape the theories developed to interpret and explain the phenomena they describe' (Gooding's emphasis).[4] These considerations are central to discussions of literature and electricity, as core elements that enabled scientists and other authors to grasp and convey concepts about the phenomenon.

Circulating within the nexus of scientific, practitioner and non-specialist interests were multiple understandings, perceptions and misconceptions of the phenomenon, which constantly intermingled, rather than existing as distinct or dual narratives. Writing about electricity was, itself, a form of exploration and experimentation, and inseparable from the innate fluidity of the period's forms of knowledge. The newly diverse authorship of the nineteenth century prompted interpretations of electricity that incorporated science, yet also extended and altered its visionary scope. The tensions of epistemology and authority, cast up by travelling from pre-nineteenth-century traditions of natural philosophy into the modern science of physics, were a further significant feature of early electrical practice.[5] However, writings about electricity complicated the apparently inexorable movement towards the scientific authority of quantification, precision and control, in seeking to balance the new features of physics with older ideas, drawn from earlier associations with the phenomenon. The importance of the human body has been regularly acknowledged and explored by scholars, as a means of experiencing and engaging with ideas about electricity.[6] The present work

---

[2]    Geoffrey Cantor and M.J.S. Hodge, *Conceptions of Ether: Studies in the History of Ether Theories, 1740–1900* (Cambridge: Cambridge University Press 1981); Peter M. Harman, *Energy, Force, and Matter: The Conceptual Development of Nineteenth-Century Physics* (Cambridge: Cambridge University Press, 1982); David Gooding, Trevor Pinch and Simon Schaffer, eds, *The Uses of Experiment: Studies in the Natural Sciences* (Cambridge: Cambridge University Press, 1989).

[3]    Geoffrey Cantor, 'The Rhetoric of Experiment', in Gooding et al. (1989), 160–61.

[4]    David Gooding, '"Magnetic Curves" and the Magnetic Field: Experimentation and Representation in the History of a Theory', in Gooding et al. (1989), 192.

[5]    Iwan Rhys Morus, *When Physics Became King* (Chicago: University of Chicago Press, 2005).

[6]    For example, Carolyn Marvin, *When Old Technologies were New: Thinking about Electric Communication in the Late Nineteenth Century* (Oxford: Oxford University Press, 1988); Herbert Sussman, *Victorian Technology: Invention, Innovation, and the Rise of the*

aims to extend the scope of those enquiries, by considering how writings about electricity also queried the nature of corporeality, perception and interpretation.

The potential, scope and purposes of electrical power engaged vastly different audiences, partly in terms of the social status of practitioners but also because of their widely divergent, experimental approaches and aims. Neither did scientists write only about technicalities; their work was often performative too, aiming to convey and communicate the experience – not just the findings – of science, of discovery, and of electricity. Whether written by scientists or lay practitioners, representations of electricity frequently constituted rhetorical performances. Revealed in scientists' notes, correspondence and poems, and writings by ostensibly non-scientific authors, are similarities and differences, both of which contributed to the cultural integration of ideas about electricity. Electrical sciences did not exist in a vacuum, and the contexts within which they emerged did not act only as background; they were key components in how electrical phenomena were investigated.[7] Writings about electricity enable us to see the indispensable involvement of publication cultures and literary ideas, together with social and commercial changes, and how they worked alongside evolving apprehensions of the self and the physical world.

Before the nineteenth century, public awareness of electricity arose in response to the visual and sensational character of electricity, in the natural elements or through demonstrations of its effects. Interest in electricity was based on its value as a novelty and spectacle, rather than analysis or investigation of its substance and operation. Eighteenth-century parlour games such as the 'Electric Boy', pioneered in part by Francis Hauksbee at the Royal Society, demonstrated how electricity could pervade the most palpably material substance – the human body.[8] In a 1748 image of the game by physician and botanist William Watson (1715–1787), a rotating crank generates electricity that is transferred through the feet of a suspended figure; the electricity is transmitted to a girl's hand, while her other hand makes contact with feathers or small pieces of paper, attracted by the electricity passing through her. Throughout the first quarter of the nineteenth century, electricity was also widely popularised through images of Giovanni Aldini's notorious experiments, which demonstrated the effects of electricity by moving the decapitated heads of cattle and the bodies of executed criminals. These popularisations, whether unwittingly or not, established lurid and sensational associations between electricity and the

---

*Machine* (Santa Barbara: ABC-CLIO, 2009); Iwan Rhys Morus, *Shocking Bodies: Life, Death and Electricity in Victorian England* (Stroud: History Press Ltd, 2011).

[7]   My study of this relationship is informed further by: David Gooding, *Experiment and the Making of Meaning: Human Agency in Scientific Observation and Experiment* (Dordrecht: Kluwer Academic, 1990); Frank A.J.L. James, ed., *'The Common Purposes of Life': Science and Society at the Royal Institution of Great Britain* (Aldershot: Ashgate, 2002); David Knight, *The Making of Modern Science: Science, Technology, Medicine and Modernity: 1789–1914* (Cambridge: Polity, 2009).

[8]   Francis Hauksbee (1688–1763), instrument maker and lecturer on science.

body.[9] As James Secord comments, the effects of science 'seemed accessible because they were made visible', and he takes as an example John Henry Pepper's 'ghosts', in which electricity was also used to amaze credulous publics.[10] The use of the body to demonstrate electrical experiments was particularly common, for example, in Faraday's reports of making an electrical circuit between his tongue and his eyeball, to 'perceive the *sensation* upon the tongue and the *flash* before the eyes' (author's emphases).[11] The historical relationship between electricity and the body is well-documented and considerable scholarship exists about it.[12] Carolyn Marvin attributes the persistent connection between electricity and the body to the latter's role as 'the most familiar of all communicative modes ... a convenient touchstone by which to gauge, explore, and interpret the unfamiliar'.[13] In my research, electrical experimentation with the body represents an initial foundation of interest in the subject and a starting point in the transition of electrical science from sensational demonstration to abstract theory.

Early nineteenth-century representations of electricity's substance and processes also stemmed from the eighteenth century. In 1733, Charles du Fay announced his 'two fluids theory' in which electricity was proposed to consist of two electricities: the 'vitreous' and the 'resinous'.[14] He showed that substances composed of the same electricity repel one another, while substances of opposite

---

[9]    This was not necessarily Volta's intention for, as Patricia Fara points out, 'despite his renown as an inventor and experimenter, Volta regarded his theoretical research as a natural philosopher to be his greatest achievement'; see Patricia Fara, 'Alessandro Volta and the Politics of Pictures', *Endeavour*, 33:4 (2009), 126.

[10]    James A. Secord, 'Quick and Magical Shaper of Science', *Science*, 297:5587 (September 2002), 1649. John Henry Pepper (1820–1900) produced realistic ghosts through optical projections, screens and special lighting, initially in 1862 at the Royal Polytechnic Institution in London.

[11]    Michael Faraday, *Experimental Researches in Electricity*, vol. 1 (London: Richard and John Taylor, 1839), 15. *Experimental Researches* was published in 30 volumes between 1832 and 1856. The focus in the present work is predominantly on Faraday's earliest experiments and conceptualisations of electricity, which are offered in the first volume and remain relatively unchanged in subsequent editions. I am grateful to the Daubeny Library at Magdalen College Library, Oxford, for granting me access to the first editions of all the volumes.

[12]    Margaret Rowbottom and Charles Susskind, *Electricity and Medicine: History of their Interaction* (San Francisco: San Francisco Press, 1984); Meryl R. Gersh, *Electrotherapy in Rehabilitation* (Philadelphia: Davis, 1992); Autumn Stanley, *Mothers and Daughters of Invention: Notes for a Revised History of Technology* (Metuchen and London: Scarecrow, 1993); Paola Bertucci and Giuliano Pancaldi, *Electric Bodies: Episodes in the History of Medical Electricity* (Bologna: Università di Bologna, 2001); Iwan Rhys Morus, *Bodies/ Machines* (Oxford: Berg, 2002); Jason Rudy, *Electric Meters: Victorian Physiological Poetics* (Athens, OH: Ohio University Press, 2009); Morus (2011).

[13]    Marvin (1988), 109.

[14]    Charles François de Cisternay du Fay (1698–1739), 'Two Kinds of Electrical Fluid: Vitreous and Resinous', *Philosophical Transactions of the Royal Society* 38 (1733), cited in

types were attracted, providing the basis of later concepts of negative and positive charge.[15] Beyond this basic distinction, electricity, light, heat and magnetism were grouped together under the term 'imponderables', meaning that they appeared to have no quantifiable substance and yet their essences were so different, they could not interact with each other.[16] Although the term 'imponderable' had a quite specific meaning in physics, it underwent a subtle semantic shift between the late-eighteenth and mid-nineteenth centuries.[17] It described elements thought to have no perceptible weight, for example, in George Adams's description of 'phlogiston, a substance as imponderable as fire'.[18] Thereafter, it referred to light, heat and electricity and later to the luminiferous 'ether', as when John Imison observed that 'light ... is reckoned among the imponderable bodies'.[19] However, its figurative meaning was also increasingly apparent, so that it came to refer to aspects of the natural world perceived as incalculable and unthinkable, such as in Herbert Mayo's comment that 'Mind, like electricity, is an imponderable force'.[20] This conceptual transitioning implies a gradual recognition that the physical world could not be understood in purely quantitative terms, and that describing it inevitably involved an element of uncertainty and abstraction.

The relationship between conceptualising electricity's properties and other responses to it can be understood best by reviewing how theoretical approaches emerged, and how old and new concepts competed with each other as representations. Theories of electricity's imponderability were seriously challenged from 1800, when Volta demonstrated that electricity could travel between one

---

Joseph F. Keithley, *The Story of Electrical and Magnetic Measurements: From 500 B.C. to the 1940s* (New York: Institute of Electrical and Electronic Engineers, Inc., 1999), 19.

[15]   Electrical attraction and repulsion were demonstrated as an electrostatic force by Charles-Augustin de Coulomb (1736–1806), known as Coulomb's Law (1783). Emmanuel Kant proposed that attraction and repulsion are two forces, from which electricity is produced; see discussion of Emmanuel Kant's *Metaphysical Foundations* (1786) by Michael Friedman in *Kant and the Exact Sciences* (Cambridge, MA: Harvard University Press, 1992), 236.

[16]   L. Pearce Williams, 'The Physical Sciences in the First Half of the Nineteenth Century: Problems and Sources', *History of Science*, 1 (1962), 3.

[17]   The term 'physics', referring to properties of non-living matter and energy, is relatively recent; its earliest British use in this sense appears to have been in 1834 by Mary Somerville (OED) but, as late as 1871, William Thomson called his textbook on physics a *Treatise on Philosophy*; see Greg Myers, 'Nineteenth-Century Popularizations of Thermodynamics and the Rhetoric of Social Prophecy', *Victorian Studies*, 29:1 (Autumn 1985), 35–66. Peter Harman points out that, by the early nineteenth century, the term was being used 'to denote the study of mechanics, electricity and optics, employing a mathematical and experimental methodology'; see Harman (1982), 1.

[18]   *Nat. and Exp. Philos.*, I. xi. 449, 1794 (OED).

[19]   *Science and Art*, 2:33, 1822 (OED).

[20]   *Pop. Superst.*, 2nd edn, 70, 1851 (OED).

metal plate and another in his voltaic pile.[21] William Nicholson and Anthony Carlisle demonstrated further that imponderables could indeed interact because the current from the pile could decompose water, breaking it down through heat into its constituents of hydrogen and oxygen, as steam.[22] Despite the revolutionary nature of such developments, their public dissemination was limited; as one contemporary writer remarks about his own pamphlet on electricity in 1838, it was 'contrary to the design of this work to enter into any lengthened examination of these theories', on the basis that it might 'bewilder the reader by leading him through the maze of speculative theory'.[23] Experiments and demonstrations could be carried out by simply following instructions, without necessarily achieving or, indeed, seeking detailed understanding of how electricity worked.

By the 1830s, a degree of exasperation is evident over the prevailing mysteries of electricity in attempts to write explanations of it; a columnist in 1835 remarks, for example, that 'we are ignorant of how it is roused, or of the manner of its existence in bodies' and, three years later, the conclusion is still that 'in what it consists or how it is constituted, are questions too difficult for us to solve. We do not even know whether electricity is material or not'.[24] Superficial, spectacular and aesthetic representations of electricity continued to exist, but they did so alongside a fascination with the material nature of electricity and the extent to which it actually existed, questions that required the deeper analysis and greater precision of scientific investigation. It has been suggested that leading nineteenth-century scientists were an 'elite' body, which 'formed the apex of a new intellectual class'.[25] The structure both within and below that apex was less clear; as Graeme Gooday suggests, there was 'neither a simple hierarchy nor neat demarcation between popular and technical treatment' of electricity, although, as he points out, 'different groups had distinctively different concerns'.[26] Inevitably, distinctions existed between the mere spectacle of electricity and the specialist knowledge by which it could be systematically investigated. Yet the two activities were not entirely separate. Understanding electricity did not deny the attractions of the

---

[21]   Count Alessandro Giuseppe Antonio Anastasio Volta (1745–1827). The invention was termed the 'pile' by William Nicholson; see Giuliano Pancaldi, *Volta: Science and Culture in the Age of Enlightenment* (Princeton: Princeton University Press, 2005), 247.

[22]   William Nicholson (1753–1815); Sir Anthony Carlisle, FRCS, FRS (1768–1840); Antoine-Laurent de Lavoisier (1743–1794) discovered oxygen in 1778 and hydrogen in 1783.

[23]   G.H. Bachhoffner, *A Popular Treatise on Voltaic Electricity and Electro-Magnetism* (London: Simpkin and Marshall, 1838), 8, 20.

[24]   'Electricity', *Saturday Magazine*, 6:169 (21 February 1835), 68; 'Electricity', *Saturday Magazine*, 13:399 (22 September 1838), 111.

[25]   S.S. Schweber, 'Scientists as Intellectuals: The Early Victorians', in *Victorian Science and Victorian Values: Literary Perspectives*, ed. James Paradis and Thomas Postlewait (New Brunswick: Rutgers University Press, 1985), 2.

[26]   Graeme Gooday, *Domesticating Electricity: Technology, Uncertainty and Gender, 1880–1914* (London: Pickering and Chatto, 2008), 233, 37.

phenomenon; quite the opposite. Analysing its complex and abstract matter meant delving into invisible and exciting mysteries that had never been modelled before.

To appreciate literary responses to electricity fully, it is useful and even necessary to understand, at the outset, how it came to be viewed as a single phenomenon and, just as importantly, the alternative methods of representation with which verbal descriptions competed, particularly mathematics. The unification of electricity as a single concept was achieved through a combination of theoretical developments in the first half of the century.[27] By 1850, the 'imponderables' of light, heat and magnetism were largely unified under the concept of 'energy'.[28] As Peter Harman explains, 'the physical problems of light, heat and electricity were conceptualised in a way that made them amenable to mathematical analysis and thereby fostered the unification of physics'.[29] The revelation that electricity was not 'imponderable' marked a turning point in which, historians of science agree, the role of mathematics was fundamental.[30] An important feature of the new electrical vision, as Graeme Gooday argues, was a marked and growing emphasis on measuring, accuracy and numbers, as the correct way to understand and represent electricity.[31] Mathematics could extend the representative potential of verbal and visual techniques and, simultaneously, provide the precision essential for conceptualising increasingly complex concepts.[32]

Mathematics was not always intimately associated with electricity; however, Newton's *Principia Mathematica* (1687) effectively 'set the standard for the mathematical understanding of nature'.[33] In the nineteenth century, electrical science (and the energy sciences more broadly) were increasingly formalised by mathematics, creating a tone in the physical sciences that sometimes persists today. The modern physicist Wayne Saslow suggests, for example, that 'physics is a set of facts about the real world, and a coherent set of relationships between these facts, whose natural expression is through the language of mathematics'.[34] The extent to which mathematics was a 'natural' expression is debatable, and,

---

[27]  By figures such as Joseph Fourier, Hans Christian Oersted, Michael Faraday, William Thomson, James Prescott Joule, Hermann von Helmholtz and James Clerk Maxwell.

[28]  Johann Bernhard Stallo, *The Concepts and Theories of Modern Physics* (1881), quoted in Harman (1982), 1.

[29]  Harman (1982), 4.

[30]  See Cantor and Hodge (1981); Harman (1982); Morus (2005).

[31]  Graeme J.N. Gooday, *The Morals of Measurement: Accuracy, Irony and Trust in Late Victorian Electrical Practice* (Cambridge: Cambridge University Press, 2004).

[32]  The foundation is explored further by later scientists, such as William Thomson (1824–1907), George Gabriel Stokes (1819–1903), and Peter Guthrie Tait (1831–1901).

[33]  Iwan Rhys Morus, *Frankenstein's Children: Electricity, Exhibition, and Experiment in Early-Nineteenth-Century London* (Princeton: Princeton University Press, 1998), 23.

[34]  Wayne M. Saslow, *Electricity, Magnetism, and Light* (Amsterdam: Academic Press Elsvier Science, 2002), xiii.

as Saslow later concedes, both Faraday and Maxwell also 'thought visually'.[35] Advanced understandings of electromagnetism were launched by the successful magnetisation of wires by Hans Christian Oersted (1777–1851) in 1820 and spurred on by the advent of practical developments, such as batteries and voltaic piles.[36] Advances such as these were not necessarily reliant upon mathematics, if at all, but they did demonstrate crucial relationships between electricity, magnetism and matter, which, at least initially, could only be described using images and verbal descriptions.

A further decisive moment in the representation of electricity was the creation by André-Marie Ampère of a unit by which it could be measured with reasonable accuracy.[37] Ampère's contributions are well-known but the work of many others was also significant, such as Siméon-Denis Poisson, George Green and William Rowan Hamilton, which are reviewed briefly here before focusing more closely on the work of Michael Faraday and James Clerk Maxwell. In 1826, a summarised translation of Siméon-Denis Poisson's (1781–1840) attempts to mathematise magnetic bodies appeared in the *Quarterly Journal of Science*; despite its relative obscurity, Poisson's work represented one of the earliest attempts to apply mathematical theory to electrical science.[38] Similarly obscure was George Green's *An Essay on the Mathematical Analysis of Electricity and Magnetism* (1828), in which he attempted to determine the 'density' of electricity and magnetism using potential functions and equations. Like Poisson, Green remained almost entirely unknown until the 1850s, when his work was finally 'recognised as a major text in potential theory'.[39] The Irish mathematician William Rowan Hamilton (1805–1865) also made what has been described as 'the earliest significant nineteenth-century contribution to mathematical physics', an involvement that began in the 1830s when he devised a system of first-order partial differential equations involving a single function; these provided a central foundation for later laws of physics involving gravitation, optics and dynamics, as well as electricity.[40] By 1843, Hamilton's place in the mathematical history of the energy sciences was secured by his invention of the algebraic quaternion, which allowed for much greater concision in using the equations of earlier European eighteenth-century

---

[35]    Saslow (2002).

[36]    Crosbie Smith, *The Science of Energy: A Cultural History of Energy Physics in Victorian Britain* (London: Athlone, 1998).

[37]    The electrical unit of measurement, the ampere, was named after Ampère (1775–1836).

[38]    I. Grattan-Guinness, ed., *Landmark Writings in Western Mathematics 1640–1940* (Amsterdam: Elsevier B.V., 2005), 404.

[39]    Grattan-Guinness (2005), 403.

[40]    Carl Benjamin Boyer, *A History of Mathematics*; revised by Uta C. Merzbach, 2nd edn (New York: Wiley, 1989), 622.

theorists, such as Joseph-Louis Lagrange (1736–1813) and Pierre Simon de Laplace (1749–1827).[41]

In approaching questions of scientific modelling from the perspective of literary studies, my central interest is in the parallels and divergences between the mathematical modelling, towards which electrical science appeared to move in the first half of the century, and writing, as a scientific modelling technique. The cognitive scientist Nancy Nersessian confirms that scientists' modelling practices are not 'inessential aids' but, rather, ways of reasoning and understanding, creating and using theories.[42] My research focuses primarily on written responses and techniques, but I also aim to show that correlations can exist between ostensibly different methodologies. Just as literary techniques like metaphor and imagery often rely upon visual appreciation, so mathematics and poetry share a fundamentally symbolic nature. As Barbara Maria Stafford remarks, true interdisciplinarity is 'grounded in the acknowledgement that perception (*aisthesis*) is a significant form of knowledge (*episteme*), perhaps even the constitutive form'.[43] My research does not aim to impose intellectual hierarchies on verbal, mathematical and visual forms of knowledge; instead, it considers them to be equally vital, as the means by which nineteenth-century scientists perceived and sought to understand electricity.

## Heuristic Fictions

Speculation was a vital element of conceptualising electricity in the nineteenth century; indeed, the inherently hypothetical nature of speculation could itself be viewed as a form of fictionalising, so much so that philosophers of physics describe initially speculative theories as 'heuristic fiction'.[44] Speculation was absolutely central to Michael Faraday's investigations of electricity and his view of scientific endeavour, a feature frequently eclipsed by the importance of his discoveries.[45] Indeed, as Cantor, Gooding and James point out in their review of his life and work, 'excessive concentration on discovery distorts our understanding of both

---

[41]    Albert C. Lewis, 'Hamilton, Sir William Rowan (1805–1865)', *Oxford Dictionary of National Biography* (Oxford University Press, 2004) [www.oxforddnb.com/view/article/12148, accessed 25 January 2009].

[42]    Nancy J. Nersessian, 'Model-Based Reasoning in Conceptual Change', in *Model-Based Reasoning in Scientific Discovery*, ed. Lorenzo Magnani, Nancy J. Nersessian and Paul Thagard (New York: Kluwer Academic/Plenum, 1999), 8.

[43]    Barbara Maria Stafford, 'Visualization of Knowledge', in *Consumption and the World of Goods*, ed. John Brewer and Roy Porter (London: Routledge, 1993), 473.

[44]    Dugald Murdoch, *Niels Bohr's Philosophy of Physics* (Cambridge: Cambridge University Press, 1989), 75.

[45]    Faraday's discoveries included establishing the principle of electromagnetic induction (1821), demonstrating the generation of electricity by means of magnetism and motion (1831), and discovering the magneto-optical effect and diamagnetism (1845). See John Tyndall, *Faraday as Discoverer* (London, 1868).

scientists and the activity of science'.[46] By considering the conceptual techniques Faraday employed in responding to electricity, I aim to recover some of the processes, activities and experiences that were as much a part of his scientific contribution as his discoveries.

Faraday was the field's leading specialist but he was also a writer. Celebrated by contemporaries for his 'perfect simplicity of thought and language' in promoting awareness about the nature of electricity, he saw the study of electricity as a key aspect of society's future progress.[47] This is evident in Faraday's own comments on the extraordinary 'progress which electricity has made in the last thirty years' and his declaration that 'no branch of knowledge can afford so fine a field for discovery as this'.[48] Faraday's informal yet vibrant first-person narratives left readers in no doubt as to the importance of studying electricity and, in doing so, he provided a template for writing about electricity upon which many subsequent books were modelled. Faraday's creativity as a natural philosopher fuelled his resistance to the established mechanistic and Newtonian view of nature, as well as later mathematical approaches. By mid-century, his work was recognised as 'unquestionably among the most important that the century has produced' and, more particularly, as having 'given a direction to the ablest scientific thought of our day'.[49] Faraday's research on electricity presented for the first time a 'system of connected knowledge', and the relationships he demonstrated between electricity, magnetism, gravity, light, thermodynamics and atomic matter showed the universe to be made up of 'continuous matter possessed of kinetic energy'.[50] In the creation of this new kinetic portrayal, however, scientists were increasingly aware of the distinctions between what they were representing and how they were representing it. The gap between 'the structure of physical reality and the encompassing net of theory' was, as Peter Harman asserts, 'a theme of fundamental importance' to nineteenth-century electrical investigators.[51] It was also a breach that complicated what might, otherwise, appear to be a logical progression from mystery to mastery and it indicates the deeply speculative nature of nineteenth-century physical sciences.

Having trained under Humphry Davy, Faraday investigated electricity as an experimentalist; as he comments, 'a physical line of force may be dealt with experimentally, yet without our knowing its *intimate physical nature*' (Faraday's

---

[46]    Geoffrey Cantor, David Gooding and Frank A.J.L. James, *Faraday* (Basingstoke: Macmillan, 1991), 99.

[47]    John Hall Gladstone, 'Faraday: Review of The Life and Letters of Faraday by Dr. Bence Jones', *Nature* (17 February 1870), 403.

[48]    Faraday, Untitled, *Annals of Electricity, Magnetism and Chemistry*, 4:6 (July 1839), 9.

[49]    'Science', *Westminster Review*, 65:127 (January 1856), 254.

[50]    Cantor et al. (1991), 99; Smith (1998), 2.

[51]    Harman (1982), 9.

emphasis).[52] The shift towards an atomic and microscopic study of nature differed substantially in scale and intent from the previously passive observation of electricity's effects, making responses to electricity also experiments in modelling and communication, as much as searches for understanding. Faraday argues, however, that the 'physical character' of electricity is 'not proved' by experimentation and suggests, instead, that

> We know no more of the physical nature of the electric lines of force than we do of the magnetic lines of force; we fancy, and we form hypotheses, but unless these hypotheses are considered equally likely to be false as true, we had better not form them.[53]

Faraday's statement indicates that he neither sought to provide a Schopenhauerian 'unclouded mirror of the world' nor to follow an agenda of scientific realism, in portraying a 'true' picture of the world, 'faithful in its details'.[54] Instead, he declares that the science of electricity is based on hypotheses that should not be taken as proven or stable truths. Elsewhere, he argues that the conjectures of hypothetical speculations are 'wonderful aids in the hands of the experimentalist and mathematician' and that, as long as they are 'cautiously advanced', they are 'useful in rendering the vague idea more clear for the time', while further experiment and calculation takes place.[55] David Gooding suggests that it was Faraday's ability to use conjectural models that allowed him to move so easily between the particular phenomena of his experiments and the more general features of dynamic structures.[56] Faraday perceives speculations to be an essential part of the creative scientific process because 'they lead on, by deduction

---

[52] Michael Faraday to Carlo Matteucci, 2 November 1855, in *Correspondence of Michael Faraday, 1855–1860*, vol. 5, ed. Frank A.J.L. James (London: The Institution of Electrical Engineers, 2008), 2.

[53] James (2008).

[54] Arthur Schopenhauer, *Die Welt als Wille und Vorstellung*, quoted in Lorraine Daston and Peter Galison, *Objectivity* (New York: Zone Books, 2007), 203; Bas C. Van Fraassen, *The Scientific Image* (Oxford: Oxford University Press, 1980), 7. By 'scientific realism', I refer to the concept offered in such works as Richard Boyd's 'On the Current Status of Scientific Realism', in *The Philosophy of Science*, ed. Richard Boyd, Philip Gasper and J.D. Trout (Cambridge, MA: MIT Press, 1991); Jarrett Leplin, *A Novel Defense of Scientific Realism* (Oxford: Oxford University Press, 1997); and Stathis Psillos, *Scientific Realism: How Science Tracks Truth* (London: Routledge, 1999).

[55] Michael Faraday, 'Experimental Researches in Electricity' [1852], in *Literature and Science in the C19th Century: An Anthology*, ed. Laura Otis (Oxford: Oxford University Press, 2002), 56. Although the word 'speculation' has additional older and newer meanings (eyesight, financial ventures), its primary meaning remains the same as the sense in which Faraday uses it.

[56] David Gooding, 'From Phenomenology to Field Theory: Faraday's Visual Reasoning', *Perspectives on Science*, 14:1 (Spring 2006), 40–65.

and correction, to the discovery of new phenomena, and so cause an increase and advance of real physical truth'.[57] Faraday's use of the word 'experimental' in the title of his *Experimental Researches in Electricity*, published between 1832 and 1856, is not casual; it reflects the tentative and exploratory nature of Faraday's methodology and the wider character of nineteenth-century investigations of electricity. Faraday explicitly acknowledges, as Cantor, Gooding and James point out, that his theories are just that – 'always provisional'.[58] The 'speculative thread' of Faraday's science is an important and characteristic feature of his research on electricity, and it has been recognised as 'a crucially important influence on the physics of the late-nineteenth and early-twentieth centuries'.[59] The influence is evident, for example, in Werner Heisenberg's famous 'uncertainty principle' of 1927.[60] More recently, too, Stephen Hawking indicates the importance of speculation when he claims that, in scientific research, 'all one can do is describe what has been found to be a very good mathematical model ... and say what predictions it makes'.[61] In conceptualising and modelling complex problems, speculation is a vital component that is shared by both mathematics and verbal forms of knowledge-making.

Samuel Smiles claimed that Faraday's scientific achievements were realised not by sudden flashes of inspiration but 'by dint of mere industry and patient thinking'.[62] The image is supported by the meticulous nature of Faraday's research notes, writings that can themselves be read as forms of literature and narratives of scientific practice. Faraday indicates that he writes about his experiments and results 'not as they were obtained, but in such a manner as to give the most concise view of the whole'.[63] He writes in the first person and, in a distinctly literary manner, he presents his experiences and findings in the form of a story. Hovering alongside the apparatus and operations is Faraday the narrator who, for example, has to 'refrain (though much tempted) from offering further speculations'.[64] The man within the science peeps through the parenthetical window – 'refrained' yet ironically present.[65]

Faraday's urge to 'refrain' represents the type of self-abnegating impulse associated with nineteenth-century science discussed in recent scholarship.[66]

---

57    Faraday (1852), 56.

58    Cantor et al. (1991), 101.

59    Cantor et al. (1991), 91.

60    See David Lindley, *Uncertainty: Einstein, Heisenberg, Bohr, and the Struggle for the Soul of Science* (London: Doubleday Books, 2007).

61    Stephen W. Hawking, *The Universe in a Nutshell* (London: Bantam, 2001), 31.

62    Samuel Smiles, *Self-Help: With Illustrations of Character and Conduct* (1859; repr. London: Routledge/Thoemmes Press, 1997), 78.

63    Faraday (1839), 2.

64    Faraday (1839), 22.

65    Faraday (1839).

66    George Levine, *Dying to Know: Scientific Epistemology and Narrative in Victorian England* (Chicago: University of Chicago Press, 2002); Daston and Galison (2007).

However, Faraday also refers regularly to his hopes, his expectations and even his determination not to despair; he seeks answers to one experiment 'with great anxiety' and is 'strongly stimulated' by the 'beautiful' results of another.[67] He portrays the scientific self as a vital part of the undertaking's creativity, and historians agree that his experimentation was a 'highly reflective and a very personal activity', one that was 'directed by goals and values outside his laboratory'.[68] He is also keenly self-aware, admitting at one point that he has to guard 'with great suspicion of the influence of favourite notions over myself' and he reveals his thinking at various stages, his motivations and key developments by other scientists.[69] Rather than striving for the objectivity that 'scruples to filter out the noise that undermines certainty', Faraday documents the flawed, subjective and error-ridden nature of electrical experimentation by himself and his fellow scientists.[70] He comments excitedly on the thrilling and sensational nature of experimentation and the 'exceedingly remarkable and novel consequences' of his experiments.[71] The environment he portrays is part of, not separate from, contemporary life, and more akin to Bruno Latour's contention that 'when we go from "daily life" to scientific activity, from the man in the street to the men in the laboratory, from politics to expert opinion, we do not go from noise to quiet, from passion to reason, from heat to cold'.[72] Instead of seeking selfless impartiality, Faraday portrays himself as inseparable from the processes and interpretation of science. As the historian Theodore Porter suggests, it is important to look 'outside the autonomous development of science and examine its place in the larger world of political and economic life'.[73] In reading Faraday's research, we are presented with a highly subjective response to electricity by one of its most celebrated figures, who welcomes individuality as an essential feature of science.

The extent to which Faraday's accounts are literary presentations is evident if we consider his association with Ampère. Throughout *Experimental Researches*, Faraday regularly quoted Ampère, described his experiments and even compared their experimental failures, indicating the existence of a close relationship between the two men beyond the text. However, as Latour also suggests, 'when we approach the places where facts and machines are made, we get into the midst of controversies ... We go from controversies to fiercer controversies'.[74] Any

---

[67]   Faraday (1839), 538.

[68]   Cantor et al. (1991), 101.

[69]   Faraday (1839), 386.

[70]   Daston and Galison (2007), 17.

[71]   Michael Faraday in a letter to Richard Phillips, 20 November 1834; 'Additional Observations Respecting the Magneto-Electric Spark and Shock', *Philosophical Magazine and Journal of Science* (London and Edinburgh, December 1834).

[72]   Bruno Latour, *Science in Action* (Milton Keynes: Open University Press, 1987), 30.

[73]   Theodore M. Porter, Review: 'The Objective Self', *Victorian Studies*, 50:4 (Summer 2008), 646.

[74]   Latour (1987), 30.

affiliation between Ampère and Faraday was, effectively, a fiction; however, Faraday, ever the gentleman, declines to mention the serious flaws he had exposed in Ampère's work. The controversy between Faraday and Ampère was, indeed, about 'facts and machines' but it was also about scientific approaches and reputations.[75] Ampère's theory of electrodynamics was based on mathematical suppositions, rather than experiment and, while in certain respects 'Ampèrean currents were not incompatible with Faraday's views', Faraday did reject both Ampère's approaches and his hypothesis as a whole.[76] Interestingly, Faraday declared to William Whewell in 1835 that his denial of Ampère's ideas was based 'more on a general feeling than any thing founded on distinct objections'.[77] When he repeated Ampère's experiments, he discovered the operation of continuous electromagnetic rotation, the principle of the electric motor, a phenomenon not predicted by Ampère and not even possible within the latter's theoretical framework. Faraday appears to feel it unnecessary to publicise the episode or the flaws in Ampère's theory, perhaps because they were widely publicised elsewhere. However, the episode vindicated Faraday's 'lifelong scepticism' about the superiority of mathematical approaches, over experimental ones, for describing nature.[78] For Faraday, mathematics, theory and speculation was rooted in a combination of experimental proof and the altogether less 'scientific' response of gut instinct.

In his published notes, Faraday regularly left the laboratory and lecture room to experiment in clearly identified and familiar locations. Doing so gave readers of his research a newly participative role, for they could go to the same places to re-imagine and even sometimes watch his experiments, even if the scale of them made it unlikely that they could repeat them. In his experiment on Waterloo Bridge, for example, Faraday describes having stretched lengths of copper wire 960 feet long from the parapet of the bridge, whilst attaching other wires to metal plates on the water, in order to gauge the electromagnetic current generated by the force of the water flow.[79]

---

[75]   In Ampère's theory of electrodynamics, the luminiferous ether was a neutral and omnipresent fluid, composed of two electricities. Ampère claimed that the separation of these two electricities from the ether gave rise to all electrical phenomena, and he proposed that atoms had an unchanging and inherent electricity that circulated around their centres, attracted the opposite fluid from the surrounding neutral ether and repelled the like component until equilibrium was restored. See Trevor H. Levere, *Affinity and Matter: Elements of Chemical Philosophy 1800–1865* (Oxford: Clarendon Press, 1971), 116.

[76]   Olivier Darrigol, *Electrodynamics from Ampère to Einstein* (Oxford: Oxford University Press, 2002), 20.

[77]   Michael Faraday, quoted in Levere (1971), 120.

[78]   Frank A.J.L. James, 'Faraday, Michael (1791–1867)', *Oxford Dictionary of National Biography* (Oxford University Press, September 2004; online edition, January 2008) [http://ezproxy.ouls.ox.ac.uk:2117/view/article/9153; accessed 28 April 2010].

[79]   Faraday would certainly have known about experiments to determine the speed of electricity near London Bridge in 1747 by Folkes, Cavendish and Bevis on the Thames.

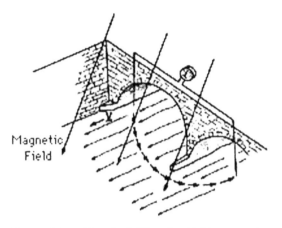

Figure 2.1    Faraday's Waterloo Bridge experiment[80]

After three days of experimenting in vain, even Faraday was forced to admit defeat, but he was also undaunted. Without pausing to discuss the failure in his report, he immediately envisioned an even grander real-world circuit, one where 'a line be imagined passing from Dover to Calais through the sea, and returning through the land beneath water to Dover'.[81] Clearly, Faraday's scientific experimentation and theorising were imaginative as well as technical; for him, science does not appear to have been limited to the laboratory and neither was the laboratory itself limited. By literally connecting electricity to London's familiar rivers and bridges, he eradicated boundaries between his science, his narrative and the world. In that sense, narrative offered Faraday an unfettered creative space, as well as an outlet for the essentially expressive and imaginative features of both his speculations and his explorations.

Representational methodologies were crucial building blocks in Faraday's approach, rather than simply vehicles for ideas or activities peripheral to scientific endeavour. His explications of electricity in *Experimental Researches* depended on verbal descriptions and labelled diagrams, rather than mathematics. He did not feel that verbal representational methods lessened the importance of precision; indeed, he stressed how essential the 'habit of forming *clear and precise ideas*' (Faraday's emphasis) is for good judgement.[82] He emphasised, too, the importance

---

Fittingly, a bronze statue of Faraday now stands at the north end of Waterloo Bridge, outside the Institution of Electrical Engineers.

[80]    Source: David P. Stern at http://www.phy6.org/earthmag/dynamos.htm; accessed 26 July 2011.

[81]    Faraday (1839), 56.

[82]    Michael Faraday, 'Observations on Mental Education' (1854), in *Michael Faraday's Mental Exercises: An Artisan Circle in Regency London*, ed. Alice Jenkins (Liverpool:

of 'clear and definite language, especially in physical matters' and of 'giving to a word its true and full, but measured meaning, that we may be able to convey our ideas clearly to the minds of others'.[83] Faraday's promotion of 'measured meaning' reflects a desire for careful, specific communications between scientists, as well as with the wider public, in the collaborative push against the boundaries of existing knowledge.

Faraday writes of using analogy 'to work out a discovery', indicating that he viewed representation as a central, rather than secondary, feature of scientific practice.[84] The importance of fiction as a characteristic of scientific innovation and representation is increasingly recognised by scholars, in what constitutes a reassessment of assumptions about fiction and fact, which is long overdue. The importance of language in scientific modelling is indicated by Robert Oppenheimer's suggestion that 'science is an immensely creative enriching experience ... full of novelty and aspiration; and it is in order to get to these that analogy is an indispensable instrument'.[85] More recently, Peter Godfrey-Smith proposes that imaginary systems are explored and used 'as the basis for an understanding of more complex real-world systems'.[86] Analogy is useful not merely to convey previously formulated information but also to discover new knowledge, particularly about phenomena such as electricity which, in the nineteenth century, had never been fully modelled in physical or conceptual terms.

Writing about electricity meant experimenting almost as much with words and images as with laboratory equipment; in the 1830s, as Alice Jenkins suggests, 'one of the crucial problems with designing a vocabulary for it was that there was as yet no universally agreed explanation of the nature of electricity'.[87] How electricity was imagined, represented and communicated was fundamental to scientific advancement, prompted in part by what has been described as the 1840s 'crises of vocabulary'.[88] However, as George Levine points out, 'fully to understand the "science" ... one must recognise how the language contributed to it, evoked

---

Liverpool University Press, 2008), 206.

[83]    Jenkins (2008), 207.

[84]    Letter from Faraday to Christian Friedrich Schönbein, 13 November 1845, in *The Letters of Faraday and Schoenbein, 1836–1862*, ed. Georg W.A. Kahlbaum and Francis Vernon Darbishire (London: Williams and Norgate, 1899), 149.

[85]    Robert Oppenheimer, addressing the American Psychological Association in 1955, quoted by Jamie Cohen-Cole in 'The Reflexivity of Cognitive Science: The Scientist as Model of Human Nature', *History of the Human Sciences*, 18 (November 2005), 118.

[86]    Peter Godfrey-Smith, 'Models and Fictions in Science', *Philosophical Studies*, 143:1 (March 2009), 102.

[87]    Alice Jenkins, *Space and the 'March of Mind': Literature and the Physical Sciences in Britain, 1815–1850* (Oxford: Oxford University Press, 2007), 126.

[88]    J. Martin and R. Harré, 'Metaphor in Science', in *Metaphor: Problems and Perspectives*, ed. D.S. Miall (Brighton: Harvester Press, 1982), 96.

resistances, entailed compliance'.[89] Analogies such as 'lines' of magnetic force allowed increasingly fine distinctions to be made in what Frank James describes as the radical 'reconceptualization of the nature of electricity'.[90] Yet no matter how expert the metaphor or elaborate the description, language appeared to be an inadequate tool for describing electricity's qualities, so that contemporary writers expressed reservations as early as 1834 about the worryingly 'loose and indefinite' first principles of physical science.[91] The rapid development of electrical studies created further instability in communication, in the proliferation of new terms. In consultation with William Whewell, Faraday introduced the words 'anode', 'cathode', 'anions', 'cations' and 'ions', and, with William Nicholl, 'electrolyte' and 'electrolyzed'.[92] The new terms provided a specific yet usable rhetoric of electricity, so that their introduction constituted both 'a heuristic device and a tool of persuasion', as well as a vital component in understanding relationships between nineteenth-century science and wider culture.[93] Whewell, however, expressed concern that notation, nomenclature and complex numbering might actually be 'hindrances to the progress of science' and even 'retard its progress'.[94] Ironically, Faraday claimed in reply to have overcome the 'hot objections' to his first public use of the word 'ion' by holding up what he refers to as 'the shield' of Whewell's authority.[95] The accelerated conceptualisation of electrical concepts also resulted in the new deployment of familiar concepts, such as 'currents' and 'fields'.[96] It has been suggested that the use of these terms may well be 'without the intention of a point-by-point comparison'.[97] However, the term 'current' certainly appears to rely heavily on early conceptions of electricity as an imponderable fluid. The word 'field', meanwhile, allows Faraday to convey the idea of an area

---

[89] George Levine, Preface to 3rd edn, Gillian Beer, *Darwin's Plots: Evolutionary Narrative in Darwin, George Eliot and Nineteenth-Century Fiction* (Cambridge: Cambridge University Press, 2009), xii.

[90] James (2004).

[91] Adam and Charles Black (Firm), Preface, *An Attempt to Simplify the Theories of Electricity and Light* (Edinburgh: Adam and Charles Black, 1834).

[92] Richard Yeo, *Defining Science: William Whewell, Natural Knowledge and Public Debate in Early Victorian Britain* (Cambridge: Cambridge University Press, 1993), 3.

[93] Jenkins (2007), 126.

[94] Letter from William Whewell to Michael Faraday, quoted in *The Philosopher's Tree: A Selection of Michael Faraday's Writings*, ed. Peter Day (Bristol: Institute of Physics, 1999), 98.

[95] Letter from Michael Faraday to William Whewell, in Day (1999), 100.

[96] 'Current': 1842 Grove, *Corr. Phys. Forces* 48 'From the manner in which the peculiar force called electricity is seemingly transmitted through certain bodies ... the term current is commonly used to denote its apparent progress'. 'Field': 1845 Faraday, *Diary*, 10 November (1933) IV. 331 'Wrought with bodies between the great poles, i.e. in the magnetic field, as to their motions under the influence of magnetic force' (OED).

[97] Martin and Harré (1982), 100.

of activity, of attention, and of space, as well as a region filled with lines of electric or magnetic force.[98]

Faraday discussed with James Clerk Maxwell, too, how best to communicate about electricity in 1857, asking (somewhat plaintively) whether after mathematical conclusions had been reached, they might not be expressed in equally clear 'common language'. While he could manage to interpret Maxwell's notations himself, Faraday felt they were slower and more unwieldy than plain descriptions and all the more so for those with less knowledge. He proposed to Maxwell that 'translating them out of their hieroglyphics' would be 'a great boon' because more people might then be able to 'work upon them by experiment'.[99] The notation and symbols of mathematics offered an alternative and more precise form of representation, but they were problematic. Even mathematicians shared concern about specialist knowledge being too narrow, for example, when the mathematician Charles Babbage bemoaned the 'ignorance' of university students educated solely in mathematics and the classics.[100] For Faraday, the practicality of experiments worked in tandem with representation, a relationship recognised by David Gooding when he suggests that 'experimentation has a constructive, inventive aspect', which 'plays an important enabling role in the presentation, interpretation and subsequent critical scrutiny of experiments'.[101] Faraday's discussions with Whewell and Maxwell about scientific representation reflect his continuing interest that concepts should be conveyed in a 'popular useful working state', as well as in more specialist forms. It is no coincidence that Faraday raised the issue with Maxwell; it was Maxwell's equations that would simultaneously advance the sciences of electricity and yet distance them almost irredeemably from non-specialist audiences.

## Fiction, Analogy and Mathematics

James Clerk Maxwell's reputation stems almost as much from his mathematisation of Faraday's work as from his own endeavours. By the time Maxwell published his *Treatise on Electricity and Magnetism* in 1873, as Alice Jenkins indicates, 'the language of field theory had shifted irrevocably away from words'.[102] In the face of electricity's abstraction, verbal description and imagery was forced to give way, at least in part, to a revolution of quantification and mathematics, albeit slow and

---

[98]    Cantor et al. (1991), 77.
[99]    Letter from Michael Faraday to James Clerk Maxwell, in Day (1999), 102–3.
[100]    Charles Babbage, *Reflections on the Decline of Science in England* (London: B. Fellowes and J. Booth, 1830), 3.
[101]    David Gooding, 'History in the Laboratory: Can We Tell What Really Went On?', in *The Development of the Laboratory: Essays on the Place of Experiment in Industrial Civilisation*, ed. Frank A.J.L. James (London: Macmillan, 1989), 66.
[102]    Jenkins (2007), 17.

impartial; the 'existing paradigm' had, in Thomas Kuhn's phrasing, 'ceased to function adequately in the exploration of an aspect of nature to which the paradigm itself had previously led the way'.[103] Faraday's experimental conclusions provided the basis for Maxwell's concept of electromagnetic radiation and field equations; however, to advance the enquiry further, Maxwell brought to bear a new grade of 'mathematical correctness' on his speculative lines of force, taking Faraday's discussions in 'new and exciting' directions.[104] Yet the characterisation of Faraday as a practical experimenter and Maxwell as a theoretical mathematician eclipses more complex aspects of their responses to ideas about electricity, which have suffered scholarly neglect. Faraday and Maxwell were scientific *writers*, as well as practitioners. In this section, I consider how Maxwell responded to ideas about electricity by employing fictional concepts and the reservations he expressed about the popularisation of scientific knowledge. Both these considerations provide useful foundations for the final section of this chapter, which examines Maxwell's poetry as an alternative form for understanding abstract concepts.

For Maxwell, analysis may have been 'grounded in both mathematics and mechanistic analogy' but it was also not exclusively or comfortably so.[105] Perhaps surprisingly, in view of his own role as a mathematician, Maxwell shared many of the reservations held by Faraday and Whewell about the increasingly specialised terminology of electrical science. He expressed regret, for instance, that 'the present state of electrical science seems peculiarly unfavourable to speculation', largely because of the domination of advanced mathematics, and he appeared genuinely to sympathise with students who had to familiarise themselves with such a 'considerable body of the most intricate mathematics ... the mere retention of which in the memory materially interferes with further progress'.[106] Despite his own extraordinary mathematical capabilities, Maxwell did not portray himself as above these intellectual struggles. Instead, he included himself when he suggested that, in battling with the conceptual connections between real-world referents and mathematics, '*we* entirely lose sight of the phenomena to be explained' (my emphasis).[107] The obstacles to conceptualising and communicating scientific ideas about electricity led him to remark that 'we must therefore discover some method of investigation which allows the mind at every step to lay hold of a clear physical conception'.[108] He celebrates how Faraday's and William Thompson's analogies bring 'before the mind, in a convenient and manageable form, those mathematical

[103]    Thomas Kuhn, *Scientific Revolutions* (Chicago: University of Chicago Press, 1996), 92.

[104]    Frank A.J.L. James, in James (2008), xxxix; Cantor et al. (1991), 92.

[105]    Cantor et al. (1991), 93.

[106]    James Clerk Maxwell, 'On Faraday's Lines of Force' [1855–6], in *The Scientific Papers of James Clerk Maxwell*, ed. W.D. Niven (New York: Dover Publications, Inc., 1965), 155.

[107]    Niven (1965).

[108]    Niven (1965), 156.

ideas which are necessary to the study of electricity'.[109] In his view, the connection between electrical science and accessible written communication was fundamental to its future progress.

Maxwell chose the fluid analogy as a model of electricity's operation, a surprising choice perhaps, considering the confusion that already existed about electrical fluids. The model indicates the importance of the distinctions Maxwell wished to convey, between mathematics and empirical science. Maxwell's analogies are described by Kevin Lambert as 'a new cognitive tool', which offered valuable insights into the complex array of social and religious influences involved in the making of scientific knowledge in the nineteenth century, as well as the role of analogy in Maxwell's mathematics.[110] Certainly, in combination with mathematical methods, analogies allowed a greater degree of concision and, indeed, more precision than was possible previously through experimental observation alone or accompanying verbal descriptions. Lambert's suggestion, however, that 'On Faraday's Lines of Force' was 'the result of Maxwell's attempt to think on paper with theoretical objects in a way analogous to Faraday's thinking with objects in the laboratory' risks neglecting the absolute centrality of pre-representational and mental modelling in science.[111] The imaginary and fictional stages of problem-solving were critical features of Maxwell's thinking and they took place in his mind, long before his pen ever met paper. His mathematisation of Faraday's experiments was about more than introducing expert notation; in some respects, it took concepts of electricity beyond the ordinary or 'real' world of empiricism, observation and sensation, and into previously inconceivable realms of abstraction.

What is particularly exciting about Maxwell's mathematical modelling is its use of fictional concepts to interpret and understand abstract realities. Mathematical theory involved more than reinforcing what Lambert describes as the 'strict distinction' between fact-producing experiments and theory and bringing 'order to those facts'.[112] Maxwell represents electricity as an '*imaginary* fluid', an adjective he accentuates.[113] Genter and Genter propose that 'people who think of electricity as though it were water import significant physical relationships from the domain of flowing fluids when they reason about electricity'.[114] However, Maxwell draws only partially on earlier concepts of liquid electricity, the luminiferous ether,

---

[109]    Niven (1965), 157.

[110]    Kevin Lambert, 'The Uses of Analogy: James Clerk Maxwell's "On Faraday's Lines of Force" and Early Victorian Analogical Argument', *British Journal for the History of Science*, 44:1 (March 2011), 61.

[111]    Lambert (2011), 62.

[112]    Lambert (2011).

[113]    Maxwell, in Niven (1965), 155.

[114]    Dedre Gentner and D.R. Gentner, 'Flowing Waters or Teeming Crowds: Mental Models of Electricity', in *Mental Models*, ed. D. Gentner and A.L. Stevens (Hillsdale: Lawrence Erlbaum Associates, 1983), 127.

or even fluidity as a concept. The fluid referent has only one or two relevant properties – the most obvious of which is motion. Maxwell asserts, though, that his electrical fluid is 'not even a hypothetical fluid', but, rather, a 'purely geometrical idea'; it moves like an ordinary fluid and it is similarly incompressible but, Maxwell stresses, it is 'merely a collection of imaginary properties'.[115] That phrase is crucially important, because it shows the extraordinary and intriguing sophistication of Maxwell's representation. He does not simply adopt metaphors to describe observed processes – what he describes cannot be observed. Instead, he uses an entirely invented form to describe the actual matter of electricity. Furthermore, the conceptual model he devises to suit his purposes is also a physical impossibility. Maxwell, arguably the most mathematical of electrical scientists, did not merely embrace the 'scientific imagination'; he exploited the possibilities of simultaneously imaginary yet scientific concepts. It was a narrative feat that allowed him to set out how electricity behaved, in a re-imagined fictional form. Certainly, there are few better examples of Gillian Beer's observation that 'when it is first advanced, theory is at its most fictive'.[116] Maxwell's employment of imaginary options reveals epistemological features more commonly associated with fiction, and the way in which he uses analogy illustrates the fundamentally imaginative aspects of specialist scientific research. Scientific explanations – and perhaps mathematical theories particularly – are generally perceived to consist of factual certainties, but that notion is tested also by 'imaginary' numbers (also known as 'nonsense', 'inexplicable', 'incomprehensible' and even 'impossible' numbers).[117] Maxwell's writing about electricity did not just relate to literature; it was a form of literature, in that it sought to address the core purpose of representing and interpreting reality.

Uncertainty did not denote vagueness in Maxwell's practice; he saw both speculation and clarification as central to science, for example, in his suggestion that the object of the paper 'On Physical Lines of Force' (1861) was 'to clear the way for speculation' before he went on to clarify the mechanical consequences of observed phenomena relating to magnetism and electricity.[118] His view resembles the recognition by his logical successor in physics, Albert Einstein, that since 'sense perception only gives information of this external world or of "physical reality" indirectly, we can only grasp the latter by speculative

---

[115] Maxwell, in Niven (1965), 160.

[116] Beer (1983), 3.

[117] Michael J. Crowe, 'Ten Misconceptions about Mathematics', in *History and Philosophy of Modern Mathematics*, ed. William Aspray and Philip Kitcher (Minneapolis: University of Minnesota Press, 1988), 270. 'Imaginary' numbers were termed so by Descartes, although they were originally described as 'sophistic' by their inventor Cardan. The further terms were used by Napier, Girard, Hygens and Euler respectively.

[118] Maxwell, in Niven (1965), 452.

means'.[119] However, Maxwell also demonstrated ambivalence about the wider ramifications of speculation. In his introductory lecture on experimental physics at Cambridge, he began by acknowledging how 'we are daily receiving proofs that the popularisation of scientific doctrines is producing as great an alteration in the mental state of society as the material applications of science are effecting in its outward life'.[120] There exists an implied ambiguity about the nature of that 'alteration', which he related further to the popularisation of science in the same lecture, by urging scientists to engage actively with the increasing public interest in scientific issues that he considered often arbitrary and misinformed. Rather than scientists accepting the wider deployment of ideas as a process beyond the remit of science, he claimed that scientists should consider themselves responsible for its 'diffusion and cultivation' and the 'spirit of sound criticism', based on proper examinations of evidence. Maxwell suggested that empiricism gave science its ultimate authority, and recommended that students refer to the scientific principles that were made apparent by observation and evidence.

For Maxwell, only astute explanations based on true observations could 'rescue our scientific ideas from that fake condition in which we too often leave them, buried among the other products of lazy credulity'.[121] His passionate defence of scientific authenticity demonstrates an awareness of the specialist community's responsibilities, akin to George Levine's suggestion that 'ideas live in culture not disembodied, but as actions, attitudes, assumptions [and] moral imperatives'.[122] His view of scientific expertise also indicated his increasing concern about the relationship of science to culture and its associated obligations. The significance of Maxwell and Faraday's multi-faceted and highly subjective involvement in how science was conveyed and communicated emerges primarily when we study their writings as literary responses, rather than simply as accidental by-products of some artificial scientific vacuum. Faraday and Maxwell remained to be convinced of the exclusive virtues of any one approach and argued for an active combination of approaches.

## Maxwell's Electric Poetry

Poetry offered Maxwell an established literary form by which to respond to the unobservable and abstract nature of electricity. Contemporary science and poetry did not always make for easy bedfellows; contemporary writers such as Tennyson asserted that 'science and poetry "feel" Nature in different ways' and that 'they

---

[119]    Albert Einstein, quoted by Gillian Beer, *Open Fields: Science in Cultural Encounter*, 2nd edn (Oxford: Oxford University Press, 2006), 163.
[120]    Beer (2006), 242.
[121]    Beer (2006), 243.
[122]    Levine (2002), 8.

have different "dreams" of Nature'.[123] Nevertheless, as Jason Rudy maintains, between Victorian poetry and physiology, there also existed a 'curious, persistent interplay through the nineteenth century between poetry and electricity'.[124] Just a few of Maxwell's poems relate to electricity, but those that do offer interesting unifications of literary and scientific techniques.[125]

What appears to be scientific poetry is not always genuinely so. As Michael Whitworth argues, 'scientific facts in literary texts need to be understood primarily as a rhetorical ploy' because 'the literary context evacuates them of their content'.[126] Maxwell's satire on differential calculus, 'A Problem of Dynamics', is distinctive because it uses the notation and processes of a real experiment, offering science a status that is more than metaphorical, rather than giving only an appearance of science.[127] As Gentner and Gentner suggest, 'a mathematical model predicts a small number of relations which are well-specified enough and systematic enough to be concatenated into long chains of prediction'.[128] The study of electricity is essentially the investigation of matter and predictions about how it behaves when subjected to various forces. Dynamics, therefore, are central. Maxwell declares that the 'the aim of physical science is to observe and interpret natural phenomena', a purpose that corresponds closely to the main occupation of poetry.[129] 'A Problem of Dynamics' portrays an ostensibly simple experiment, in which a horizontal chain is pulled from one end and the curve changes shape, yet the poem also illustrates the invisible forces and tensions involved. Maxwell deploys authentic mathematical formulae and terminology to explore very real scientific interests and, instead of subordinating the scientific to the poetic, he makes them equal. It is a reassignment that signals the poetic composition by a practising scientist and the increasing status and authority of scientific endeavour at the time.

The reading of such a technical poem is doubtless enhanced by understanding the relevant mathematical concepts, but a comprehensive explication of these would go considerably beyond the scope of this study. The full text of the poem is provided in an appendix because, although I do not analyse it in depth here, it has particular interest as an expression of contemporary anxieties about how to convey complex scientific information. Using a combination of words, numbers

---

[123]    Daniel Brown, 'Victorian Poetry and Science', in *The Cambridge Companion to Victorian Poetry*, ed. Joseph Bristow (Cambridge: Cambridge University Press, 2000), 141.

[124]    Rudy (2009), 3.

[125]    Lewis Campbell and William Garnett, *The Life of James Clerk Maxwell, with a Selection from His Correspondence and Occasional Writings and a Sketch of His Contributions to Science* (London: Macmillan and Co., 1884).

[126]    Michael Whitworth, *Einstein's Wake: Relativity, Metaphor and Modernist Literature* (Oxford: Oxford University Press, 2001), 3.

[127]    James Clerk Maxwell, 'A Problem of Dynamics', 19 February 1854, in Campbell and Garnett (1884), 625.

[128]    Gentner and Gentner (1983), 105.

[129]    James Clerk Maxwell, 'General Considerations Concerning Scientific Apparatus', in Niven (1965), 505.

and symbols, it articulates abstract features of electricity that might seem only possible for mathematics to convey. 'Narrative fantasy' was once considered to have 'nothing to do with the mathematical logic of pure science', as James Secord reminds us; however, the poem acts as a simultaneously rhetorical, fantasy-based and mathematical vehicle.[130] To model abstract concepts accurately, contemporary scientists were forced to develop mathematical methods such as vector calculus, which aimed to describe events in nature and how they worked. In that sense, experimental and theoretical types of knowledge may well be closer than they appear, as David Gooding proposes.[131] Indeed, Gooding emphasises the 'seamless' nature of what he describes as the 'web of practical, intellectual and social interactions that made up the scientific culture in which Faraday thrived'.[132] Maxwell's equations symbolised an extension of that web and the progression of science beyond physical and visual modelling. His poetry on subjects related to electricity amalgamates the two seemingly opposed methods of abstract mathematics and verbal imagery and presents them as equally valid.

The scientific 'problem' addressed by Maxwell's poem is resolved by the 'chain' of interrelated equations incorporated in the text and an accompanying graph.[133] The dilemma remains, that any detailed discussion of the poem's content is hampered by the hurdle of verbalising its content. What can be discussed instead is the relationship between the issue of representing scientific concepts and literary analysis. Scientific descriptions and imagery, like literary analysis, often rely on uncertainties and less-than-precise devices, shifting real-world referents and unreliable associations. As a text, 'A Problem of Dynamics' represents not just scientific theory but also a stage of intellectual development when information about electricity became so advanced that it almost defied verbal expression. Rather than ignoring these concerns, Maxwell uses poetry – that most keenly 'literary' of vehicles – to communicate phenomenological and theoretical essences. It is exactly the complexity of Maxwell's subject that makes poetry such a suitable means of expression; as Tyndall suggests, 'our difficulty is not with the quality of the problem, but with its *complexity*' (Tyndall's emphasis).[134] Just as mathematical symbols are a method of shorthand, the condensed poetic form can signify elusive nuances, which ostensibly lie beyond verbal description. Poetry can employ a consistency of form and brevity that also provides unexpectedly clear scientific explications; as Alan Rauch argues, the highly structured forms of poetry make

---

[130]    James Secord, *Victorian Sensation: The Extraordinary Publication, Reception, and Secret Authorship of Vestiges of the Natural History of Creation* (Chicago: University of Chicago Press, 2000), 203.

[131]    Gooding '"Magnetic Curves"' (1989), 184.

[132]    Gooding et al. (1989).

[133]    The editors, Campbell and Garnett, do not say whether the accompanying graph was devised by Maxwell.

[134]    John Tyndall, *Fragments of Science for Unscientific People: A Series of Detached Essays, Lectures and Reviews* (New York: D. Appleton and Company, 1871), 118.

it both 'scientific' and 'empathetic to the rule-governed practices of scientific enquiry'.[135] Maxwell's execution of a complex mathematical study in just 64 lines of deceptively naïve, rhyming couplets demonstrates the correspondence between the two forms. At the same time, the poetic form reveals additional imaginative possibilities, which might have remained obscure otherwise. Tumbling cascades of tightly-packed syllables carry the incessant mesmeric quality of a scientific chant, blending poetry with science and the technical with the artistic. Instead of science simply revealing nature, poetry exposes the aesthetic appeal that exists within the science.

The operation of physical forces is a central feature of electricity, and literary techniques were indispensable in understanding and explaining them. As Robert Crawford notes in his study of contemporary science and poetry, 'science itself is often underwritten by the formulations and imaginative structures developed and articulated by poets'.[136] In 'Recollections of Dreamland' (1856), it is precisely the formulations and imaginative structures of science that Maxwell explores, as a poet-scientist. In the same year, Maxwell was re-imagining Faraday's 'lines of force' as an 'imaginary fluid', which moved in thin tubes as a continuous, incompressible liquid.[137] When Maxwell asks whether the mind itself might be a simultaneously liquid and electrical medium, with dreams as 'empty bubbles, floating upwards through the current of the mind?' (l.47), he conveys the sense in which Faraday's lines were visual yet still essentially imaginary representations. Multiple levels of meaning and perspective are evident in the literary-cum-scientific nature of Maxwell's poetic imagination, offering a simultaneity of reading levels that confounds the stereotypes of both discourses, in which neither is classified as more creative, imaginative or multi-faceted than the other.

Barri Gold claims in *ThermoPoetics* that Maxwell's 'equations of electricity and magnetism are rather more elegant and timeless than his occasional verse', and that 'the two together suggest how analogous, intertwined, and mutually productive, poetry and physics may be'.[138] Maxwell enjoys playing with the electrical current analogy, adopting it, for example, as a technological metaphor for romance when a love-sick telegraph clerk exclaims in the jokingly entitled 'Valentine by a ♂ Telegraph Clerk to a ♀ Telegraph Clerk': 'O tell me, when along the line/From my full heart the message flows,/What currents are induced in thine?' (ll.1–3).[139] The metaphor resembles the speaker of Tennyson's 'Locksley

---

[135]    Alan Rauch, 'Poetry and Science', in *A Companion to Victorian Poetry*, ed. Richard Cronin, Alison Chapman and Antony H. Harrison (Oxford: Blackwell, 2002), 475.

[136]    Robert Crawford, ed., *Contemporary Poetry and Contemporary Science* (Oxford: Oxford University Press, 2006), 3.

[137]    Maxwell, 'On Faraday's Lines of Force' (1965), 155.

[138]    Barri J. Gold, *ThermoPoetics: Energy in Victorian Literature and Science* (Cambridge, MA: MIT Press, 2010), 15.

[139]    James Clerk Maxwell, 'Valentine by a ♂ Telegraph Clerk to a ♀ Telegraph Clerk', in Campbell and Garnett (1884), 320.

Hall' (1842), who declares to Amy that 'all the current of my being sets to thee'.[140] The metaphors refer to more than sentimental whimsy or even to fluidity; interlocking circles of force had been imagined previously by Faraday, in what M. Norton Wise describes as 'the mutual embrace of electricity and magnetism'.[141] In a later satire on Tyndall's materialism, Maxwell's speaker describes how 'liquid stars their watery rays/Shoot through the solid crystal'.[142] Elsewhere, too, light is imaged as a fluid:

> The lamp-light falls on blackened walls,
> And streams through narrow perforations,
> The long beam trails o'er pasteboard scales,
> With slow-decaying oscillations.
> Flow, current, flow, set the quick light-spot flying,
> Flow current, answer light-spot, flashing, quivering, dying. (ll.1–6)[143]

Yet Maxwell was all too aware of how misleading the fluid analogy could be and, in the latter poem, its potential flaws are made apparent when the speaker subsequently disparages his student's superficial understanding of science, by claiming that her eyes were given 'to mirror heaven ... And not for methods of precision' (ll.15–16).

If we examine Maxwell's other occasional verses further, though, we can see that simple rhyme schemes and brevity sometimes mask what are exceptionally 'elegant and timeless' expositions of scientific complexities. 'Reflex Musings: Reflection from Various Surfaces' (1853), in particular, describes three interrelated experiments that compare and contrast some of the distinctions Maxwell was exploring scientifically.[144] Initially, the title's pairing of reflexes and musings appears straightforward but, on closer inspection, we can see that the juxtaposition mimics the intimate and complex relationship between literary and scientific forms, such as poetry and mathematics. The term 'reflex' refers to instinctive and immediate responses that are usually bodily or physical, while 'musings' are by definition mental, meditative and constructed. The two words are oxymoronic, in presenting a diametric opposition of body and mind, but 'musings' are not quite the same as thoughts; instead, they are free from the confines of rationality and logic. They emanate from similarly intuitive sources, making them relatively synonymous.

---

[140]     Alfred Lord Tennyson, 'Locksley Hall' (1842), ll.24.

[141]     M. Norton Wise, 'The Mutual Embrace of Electricity and Magnetism', *Science*, 203 (1979), 1310–18.

[142]     James Clerk Maxwell, 'A Tyndallic Ode', in Campbell and Garnett (1884), 323.

[143]     James Clerk Maxwell, 'Lectures to Women on Physical Science (I)', in Campbell and Garnett (1884), 321.

[144]     James Clerk Maxwell, 'Reflex Musings: Reflection from Various Surfaces', in Campbell and Garnett (1884), 593.

In the second part of the title, the 'reflections' are also both the same as and different from one another. The poem is about 'reflections' in two senses: firstly, the reflections of thoughts and, secondly, the operation of the scientific Law of Reflection in light, acoustics and fluids. The two halves of the poem and the unwavering ABAB/BABA rhyme scheme represent the self-contained mirroring of 'various surfaces' within each stanza, allowing Maxwell to elucidate the complex patterns of images, events, form and sound through which physical contexts are perceived.

The poem opens with what appears at first to be an incidental reference to an urban crowd. Like a scientific observer, the speaker watches the crowd's shifting formations from above.

> In the dense entangled street,
> Where the web of Trade is weaving,
> Forms unknown in crowds I meet
> Much of each and all believing;
> Each his small designs achieving
> Hurries on with restless feet,
> While, through Fancy's power deceiving,
> Self in every form I greet. (ll. 1–8)

The scene portrayed in this first stanza is highly significant. The phenomenon of 'the dense entangled street' was an innovation of nineteenth-century physical movement that accompanied urbanisation, but it is also an early literary conceptualisation of one of the most important concepts in the history of physics, the thought experiment later known as 'Maxwell's Demon', in which an imaginary figure tracks and sorts the passing molecules.[145] As Dedre Gentner and Albert L. Stevens explain, 'besides the hydraulics model, the most frequent spontaneous analogy for electricity is the moving-crowd analogy'.[146] In the comparison, electricity is imaged as fast-moving objects passing through passageways: the numbers of people represent current; how much they push indicates voltage; and the concept of resistance is introduced by a gate or, in Maxwell's proposal, a turnstile. Effectively, the moving 'forms' of the poem represent the atomic elements of Maxwell's scientific observations, scaled up, personified and made manifest in a contemporary urban scene.

---

[145]    James Clerk Maxwell, *Theory of Heat* (London, 1871). 'Maxwell's Demon' presented a theoretical challenge to the Second Law of Thermodynamics, set out in 1854 by William Thomson, 1st Baron Kelvin (1824–1907), which states that heat differences will always be resolved by irreversible heat distribution (entropy) to reach an equilibrium. For a full explanation of the ramifications of Maxwell's Demon, see Peter Harman, *The Natural Philosophy of James Clerk Maxwell* (Cambridge: Cambridge University Press, 1998), 134–44.

[146]    Gentner and Stevens (1983), 111.

Maxwell's representation of electricity through analogy does not stop there though. In the same stanza, the 'web of Trade' and the 'birches' shadow' represent abstract physical processes which, like electrical processes, exist despite their apparent lack of visible substance. The examples pursue the same goals as the mathematical formulae and equations of Maxwell's 'On Physical Lines of Force' but they use familiar natural phenomena to make the transition from observable to abstract levels. The process of economic trade cannot be understood by examining the constituent parts, of individual transactions, currencies or interest rates, just as the process by which light creates shadow involves more than identifying the tree's presence or location. Similarly, electricity cannot be understood by looking at the material components of a battery, because it is a process rather than a substance.

The nature of 'invisible' phenomena such as electricity, Maxwell's poem proposes, can only be understood by adopting a holistic perspective, achievable through the use of fictions. To achieve such a perspective demands a conceptual mediation of space, which combines two ways of organising information that Alice Jenkins suggests are quite different: the first is a 'hub-and-ray model of spatial arrangement' in which objects, ideas or information are arranged in a pattern of circles and straight lines, 'like a wheel whose spokes all lead to a central point'; the second is an 'aerial view model which organizes by flattening an array of objects into a plane surface and lifting the observer above that plane so that it is all visible from a single point of view'.[147] The speaker in 'Reflex Musings' seeks a unified yet imaginary space, in which information is arranged in the patterns described and, simultaneously, can be viewed from an aerial perspective. The 'hub and ray' images, as Jenkins indicates, made information 'accessible in a quasi-mystical, visionary way'.[148] A similarly revelatory and intuitive element is evident in Maxwell's suggestion that 'before we can count any number of things, we must pick them out of the universe, and give each of them a fictitious unity by definition'.[149] There is something of Goethe's approach to science in Maxwell's phenomenology, a recognition of 'the science of the wholeness of nature', which can only be gained through observing the interrelatedness of effects.[150] While their understandings of phenomena also differed substantially, Goethe and Maxwell were both intrigued by how light and colours were perceived and how those perceptions related to

---

[147]    Jenkins (2007), 59. Jenkins identifies the hub-and-ray metaphor in the work of Coleridge and De Quincey, the German Romantic writer Novalis and the *Naturphilosophie* movement.

[148]    Jenkins (2007), 62.

[149]    James Clerk Maxwell, 'Are there Real Analogies in Nature?', February 1856, in Campbell and Garnett (1884), 236. The primary subject of the paper is the potential relationship between scientific principles and moral laws.

[150]    Henri Bortoft, *The Wholeness of Nature: Goethe's Way toward a Science of Conscious Participation in Nature* (Morpeth: Lindisfarne Press, 1996), 330.

scientific explanation.[151] The individual observations portrayed in Maxwell's poem compare to Goethe's understanding of colours as 'the deeds of light' and that 'we may expect from them some explanation respecting light itself'.[152] Maxwell explores not just how to represent what can be seen (be it currencies, trees or battery components) but also the importance of fictions and the imagination in thinking about the complex processes by which physical phenomena are related.

The speaker in 'Reflex Musings' is integral to Maxwell's unification of observation, experimentation and processes; he represents an amalgamation of individual and collective perspectives, as well as unified dimensions of time and space. The speaker finds himself replicated in 'every form I greet', so that distinctions between the individual self and others slip away, in the passing entities' undulations of attraction and repulsion. As though to underline the mutability of physical distinctions, the second stanza describes the speaker's image alternately advancing and retreating on the well-water's surface, reflecting back and forth, just as it did earlier with the passing faces on the street. Barri Gold proposes that the poem 'explores the likeness of light and water, nicely anticipating the fluid analogy through which Maxwell eventually develops his equations of electricity and magnetism'.[153] Maxwell does, indeed, use light and water in the poem's study of electricity's invisible forces and processes; however, he also explores how the fluid analogy relates to other ostensibly non-fluid concerns, such as optics, sound and space. In the second stanza, he writes:

> Oft in yonder rocky dell
> Neath the birches' shadow seated,
> I have watched the darksome well,
> Where my stooping form, repeated,
> Now advanced and now retreated
> With the spring's alternate swell,
> Till destroyed before completed
> As the big drops grew and fell. (ll. 9–16)

The well is a crude form of 'wave tank', the shallow glass chamber commonly used to demonstrate the properties of waves in physics and engineering. We might wonder, then, why the term 'wave' is entirely absent, despite the poem's recurrent exploration of wave principles. The term was certainly available, having been used in physics from the early 1830s, and wave actions appear to

[151]　James Clerk Maxwell, *Experiments on Colour, as Perceived by the Eye, with Remarks on Colour Blindness* (Edinburgh: Neill and Company, 1855).

[152]　Johann Wolfgang von Goethe, *Theory of Color* (1810), quoted in David Seamon and Arthur Zajonc, *Goethe's Way of Science: A Phenomenology of Nature* (New York: State University of New York Press, 1998), 19.

[153]　Gold (2010), 16.

connect the various phenomena Maxwell investigates.[154] His avoidance of the term derives, perhaps, from the fact that the poem attempts to explore the differences between the phenomena of sound, light and matter, as well as their affinities. The fluid associations of waves might have introduced a level of misinterpretation, and restricted the conceptual scope of the poem's explorations. The term 'wave' represents an extension of the fluid analogy while, in fact, the speaker explores 'various' surfaces and different types of wave action. Fluid and sound waves represent just one type of wave (the longitudinal); what the poem's experiments investigate is the action of the other, invisible yet connecting forces of electromagnetic (transverse) waves.[155]

In what appears to be an oblique reference to Plato's 'Myth of the Cave', the speaker's observations of the well's visible and fluid substance are followed by an investigation of the seeming emptiness or lack of substance in shadows, hollows, caves and echoes.

> By the hollow mountain-side
> Questions strange I shout for ever,
> While the echoes far and wide
> Seem to mock my vain endeavour;
> Still I shout, for though they never
> Cast my borrowed voice aside,
> Words from empty words they sever –
> Words of Truth from words of Pride. (ll.17–24)

In a Tennysonian 'hollow echo of my own', the speaker's shout disappears into the apparently endless 'for ever' of the hollow cave – his speech 'borrowed' and disembodied. 'Words of Pride' are reduced to 'vain endeavour', exposing the meaningless futility of language and words in the face of the ephemeral and abstract concepts of matter and space, which Maxwell sought to understand and represent.[156] The echoing, alternate rhymes of the poem 'reflect' the speaker's impressions, illuminating the dark interiority of his mind for the reader and providing a route,

---

[154]    The first use of the term 'wave' to describe sound waves appears to have been by David Brewster (1781–1868), *Letters on Natural Magic: Addressed to Sir Walter Scott* (London: John Murray, 1832), 219.

[155]    In 'transverse' waves, particle movement is perpendicular to the motion of the wave motion; in longitudinal waves, particles move in parallel to the direction in which the wave moves. Sound and fluid waves are longitudinal, whereas electromagnetic waves are transverse (José M. Carcione, *Wave Fields in Real Media: Wave Propagation in Anisotropic, Anelastic and Porous Media* (Oxford: Pergamon 2001), 321–2).

[156]    Alfred Lord Tennyson, *In Memoriam* (1850), III, ll.85–8:
'And all the phantom, Nature, stands –
With all the music in her tone,
A hollow echo of my own, –
A hollow form with empty hands'.

too, for the speaker's growth of understanding. The poem shows how the natural world and the self can be explored through representation and language.

In the final stanza, the speaker gathers together his preceding but ostensibly disjointed realisations about matter, to form a final coherent conclusion.

> Yes, the faces in the crowd,
> And the wakened echoes, glancing
> From the mountains, rocky browed,
> And the lights in water dancing –
> Each, my wandering sense entrancing,
> Tells me back my thoughts aloud,
> All the joys of Truth enhancing
> Crushing all that makes me proud. (ll.25–32)

The speaker's single unqualified 'yes' (l.25) can be read as a muted 'Eureka' moment of profound understanding, which answers the earlier 'questions strange' (l.18), even as it resembles the technique employed by seventeenth-century metaphysical dialogues, as a response to an additional presence or preceding comment.[157] As Gillian Beer notes, Maxwell was familiar with Tennyson's poetry, quoting from 'The Princess' in his 1878 Rede Lecture on the telephone.[158] Empirical observation is presented in the poem as just one form of scientific perception, one that can also result in the misleading illusions of self-referential experience, in that it only 'tells me back my thoughts aloud' (l.30). Instead of presenting merely scientific observations, the poem attempts to represent and convey the actual experience of science, and the profound understanding of physical processes to which it leads. Maxwell's intention was 'to present the mathematical ideas to the mind in an embodied form … not as mere symbols, which neither convey the same idea, nor readily adapt themselves to the phenomena to be explained'.[159] It is the combination of the words on the page and the fictions those words create in the mind that allows the processes to be fully investigated, portrayed and, most importantly, understood. The poem's fictional form allows Maxwell to convey the vast potential of both empirical and abstract knowledge, whereby experimentation reaches beyond material processes and towards metaphysical self-questioning.[160]

---

[157] For example, George Herbert's 'The Bishop', sonnet 25 (1633) or John Donne's 'Canonisation' (1633).

[158] Beer (2006), 213.

[159] James Clerk Maxwell, quoted in C.C. Gillespie, *The Edge of Objectivity: An Essay in the History of Scientific Ideas* (Princeton: Princeton University Press, 1960), 370.

[160] My focus is on the literary qualities of Maxwell's scientific concepts. For specific discussion of Maxwell's Christianity in relation to his scientific theories, see Matthew Stanley, *Huxley's Church and Maxwell's Demon: From Theistic Science to Naturalistic Science* (Chicago: University of Chicago Press, 2014).

The poem's final line reveals Maxwell's profound fascination with the visual nature of fluid motion, in the reflected image of the speaker's individual self being 'destroyed before completed' once again. The individual 'big drops grew and fell', demonstrating the wave motion of concentric circles on the water's surface before they are immersed within the spring's collective swell. The ambiguity of the 'drops' allows them to be interpreted either as rain, with the water from above mingling with that from the spring below, or as tears falling from the 'stooping' speaker. Reading the drops as tears is not simple sentimentality, though. The poem portrays fluidity as a means for the individual and human matter of the internal, physical and emotional self to be unified with nature's water, as the external and natural source of life. The poem interrogates the validity of experimentation being grounded solely in seventeenth-century Baconian empiricism, yet it also refuses to reify nineteenth-century Positivist science. The pastoral and mountainous locations might suggest an association between the Romantic sublime and creativity but the sites are not idealised. Instead, in a distinctly mid-nineteenth-century manner, they represent 'laboratories' for the understanding of scientific and natural phenomena.

Maxwell's use of poetic fictions reveals the interwoven representational fabric of literature and science about electricity. At this stage in his career, Maxwell seemed to be working towards a type of 'new language', which existed somewhere between mathematical formulae and physical hypotheses.[161] Numbers were not divorced from fiction in his conceptualisation of electricity, an approach indicated by his assertion that 'there is nothing more essential to the right understanding of things than perception of the relations of number'.[162] Maxwell's poems about electrical forces illustrate the love of concision that underpinned his mathematical skill. 'Reflex Musings' is not simply science in the form of a poem; the poem is a form of science, which allows considerations and conclusions not available elsewhere. While writers such as Faraday and Maxwell sought precision in certain aspects of representing electricity, they did not always aim for singularity of meaning. As David Gooding points out, 'Faraday had to record and describe in two dimensions an interpretation of something that he imagined as happening in three dimensions'.[163] Faraday's and Maxwell's imaginary experiments intimate that scientific discourse often demands multiple and simultaneous reading levels, in which analogy and metaphor are vital constituents.

Comparing the different types of writing helps us see how scientific responses to electricity can also be simultaneously literary and multi-faceted, offering continually overlapping interactions between existing understandings, immediate experiences and future expectations. Secondly, while modern physicists propose

---

[161]     Joseph Turner, 'Maxwell's Method of Physical Analogy', *British Journal for the Philosophy of Science*, 6:23 (November 1955), 227.

[162]     James Clerk Maxwell, 'Analogies in Nature' [February, 1856] in Harman (1990), 3.

[163]     Gooding, in *The Development of the Laboratory: Essays on the Place of Experiment in Industrial Civilization*, ed. Frank A.J.L. James (London: Macmillan Press, 1989), 77.

that the twenty-first century is 'the century of complexity', the awareness of complexity as the central issue in science arises directly from nineteenth-century negotiations of complex natural phenomena such as electricity.[164] Finally, understanding relationships between literary and scientific responses demonstrates the prominent role of representation, both beyond and as part of the process of discovery.

*

The approaches to electricity examined in this chapter demonstrate that a gradual transition took place between the eighteenth and nineteenth centuries, in which the spectacle of electricity gave way to empirical investigation and abstract theory. However, the transition was repeatedly complicated by the phenomenon's outward intangibility. Electricity had to be imagined in order to be understood or represented, making fictionality an embedded feature of its perception and conceptualisation. It was electricity's characteristically elusive nature to which both scientists and fiction writers responded. Science struggled to establish and represent its substance, while fiction writers reflected on the ramifications of its allure. Both were essentially literary responses and, as Faraday's experiments and Maxwell's analogies and poems illustrate, they demanded similar techniques. Reading them alongside one another indicates how the fundamental distinctions between the two tended to break down, revealing their intrinsically interrelated nature.

---

[164] Stephen Hawking, Untitled, *San Jose Mercury News*, 23 January 2000.

# Chapter 3
# Electrical Practitioners

Popular literature tends to be associated with fiction. However, the new publication forums for non-fiction that emerged from the early nineteenth century increased interest in electricity considerably. Before this point, the illustrious and authoritative Royal Society and the Royal Institution tended to dominate scientific endeavour, through a combination of learned memberships and royal patronage. Indeed, in the early 1800s, as Richard Yeo indicates, 'books on science were scarce – most reports on experiments and discoveries were published in the transactions of scientific societies'.[1] From the 1820s, the exploration of electricity in particular underwent a unique cultural relocation that made it an increasingly popular activity, rather than the exclusive domain of scientific elites. Interpreting specialist electrical investigations provided lucrative publishing opportunities; as James Secord points out, 'the forging of the sciences as a distinctive field of enquiry was the product of a much wider contest about access to print and the audiences for knowledge'.[2] Greater publication opportunities created a newly diverse landscape for the dissemination of ideas about electricity. Popular non-fiction books and periodical writings made vital contributions to the broader social revolution of nineteenth-century scientific participation by literary means, in making information more widely available and shaping public awareness. Cheap books about science were significant factors of popularisation and, as Jonathan Topham suggests, played a major role alongside periodicals in the creation of popular science.[3] The new dynamism between authors, readers and editors in popular books significantly altered expert and wider engagements with science. Both books and periodicals emerged within the same popular market that altered radically the literary and scientific contexts of nineteenth-century reading. Indeed, the process of popularisation, as Richard Whitley points out, complicated any sense that the creation of knowledge was

---

[1]  Richard Yeo, *Defining Science: William Whewell, Natural Knowledge and Public Debate in Early Victorian Britain* (Cambridge: Cambridge University Press, 1993), 80.

[2]  James A. Secord, 'Science, Technology and Mathematics', in *The Cambridge History of the Book in Britain*, vol. 6, 1830–1914, ed. David McKitterick (Cambridge: Cambridge University Press, 1987), 445.

[3]  Jonathan Topham, 'Publishing "Popular Science" in Early Nineteenth-Century Britain', in *Science in the Marketplace: Nineteenth-Century Sites and Experiences*, ed. Aileen Fyfe and Bernard V. Lightman (Chicago: University of Chicago Press, 2007), 135–68.

only ever a matter of 'production' or 'acquisition'.[4] At the same time, specialist periodicals brought even greater diversity. The *Annals of Electricity* provided portrayals of the experimenting community, while generalist publications like *Reynolds's Miscellany* enabled readers to explore electricity as a hobby. *Punch* and *Once a Week*, meanwhile, responded to the widespread fascination with electricity by highlighting its associations with scientific and technological progress, not to mention the unexpected degrees of disorientation they prompted.

### Popularising Electricity

The perception of electricity as a 'latent, mysterious and powerful agent' continued to prevail at the beginning of the nineteenth century.[5] Gradually, it had emerged from within the broader grouping of 'imponderables' investigated in the late eighteenth-century chemistry by Humphry Davy, James Keir and Joseph Priestley, developments of machine technology by Richard Lovell Edgeworth and James Watt, and early considerations of evolution, such as Erasmus Darwin's 'The Temple of Nature'.[6] Increasingly, it was recognised as a distinct subject of scientific focus, rather than just another element of natural philosophy; however, for its power to be utilised, it needed to be properly understood and this made explication a pressing, dominant and recurrent purpose of books on the subject. Especially conducive to writings about electricity were the changes in British publishing during the 1820s. With the cost of materials and printing processes being reduced by the introduction of steam presses, refinements of lithographic processes, stereotyping, bleached wood-pulp paper and cloth bindings, the number of published works grew as dramatically as the price of books fell.[7] Explanations of electricity could be made accessible and attractive, with high-quality lithographic reproductions and pull-out illustrations. These material enhancements were accompanied by the development of institutions that sought to promote access to scientific topics, such as the Society for the Diffusion of Useful Knowledge (SDUK; 1826) and the British Association for the Advancement of Science (1831). Like the SDUK, the British Association was established to give 'a stronger impulse and a more systematic

---

[4]    Richard Whitley, 'Knowledge Producers and Knowledge Acquirers: Popularization and a Relation between Scientific Fields and their Publics', in *Expository Science: Forms and Functions of Popularisation*, ed. Terry Shinn and Richard Whitley (Dordrecht: D. Reidel, 1985), 3–28.

[5]    T.G. Gale, *Electricity, or Ethereal Fire* (Troy: Moffit and Lyon, 1802), 3.

[6]    Erasmus Darwin, *Temple of Nature, Or, The Origin of Society: A Poem, with Philosophical Notes*, ed. Martin Priestman [www.rc.umd.edu/editions/darwin_temple/; accessed 20 July 2009].

[7]    David M. Knight, 'Scientists and their Publics: Popularization of Science in the Nineteenth Century', in *The Cambridge History of Science*, vol. 5: Modern Physical and Mathematical Sciences, ed. M.J. Nye (Cambridge: Cambridge University Press, 2003), 73.

direction to scientific inquiry', alongside the aim of removing 'any disadvantages of a public kind that may impede its progress'.[8] Corresponding changes in publication climates, purposes and readerships promoted the popularisation of scientific writing still further, altering entirely not just the types of publications that were produced but, inevitably, what was available.

The earliest books about electricity were designed for children rather than adults. Jeremiah Joyce's *Scientific Dialogues* (1805) offered 'conversations' on electricity, in the pedagogical tradition of Robert and Maria Edgeworth's *The Parent's Assistant* (1796) and *Essays on Practical Education* (1801), as well as Sir Richard Phillips's *An Easy Grammar of Natural and Experimental Philosophy: For the Use of Schools* (1807).[9] The *Dialogues* were originally 'intended for the instruction and entertainment of young people' and, in a series of lessons between a tutor and two boys, it presents teachings about electricity by literary, theatrical and experiential means.[10] The volume was revised and reprinted at least 10 times before 1868, indicating just how successfully the combination of demonstration, drama and dialogue engaged nineteenth-century readers with science. The informality of Joyce's approach resembled the Goethean form of enquiry, 'attuned to the dynamic relation between objective and subjective experience'.[11] The volume's dynamic teaching methods show the boys learning about electricity by experiencing mild electrocution – harmlessly, albeit somewhat alarmingly to modern readers. Information about electricity is delivered through a realistic yet engaging narrative, with the tutor and the boys enjoying sensational displays of electrical sparks, crackling and flashing in a darkened room. Joyce portrays the delight of exploring science and technology, revealing electricity as an aesthetic, theatrical and emotional experience. He showed how, in electricity's unique aural and visual qualities, spectacle, educational study and scientific understanding could co-exist.

Even when publications were less explicitly directed towards children, their price confined circulation to wealthier readerships. Initially, Joyce's individual volumes sold for two shillings and sixpence, making them comparatively low-priced, although as Simon Eliot reminds us, for a slim educational book in a series costing 15 shillings in total, 'the price was not insignificant and restricted

---

8    'History of the British Science Association', British Science Association Homepage [www.britishscienceassociation.org/web/AboutUs/OurHistory/; accessed 2 June 2010].

9    Jeremiah Joyce, *Scientific Dialogues: Intended for the Instruction and Entertainment of Young People; in which the First Principles of Natural and Experimental Philosophy are Fully Explained* (1805; repr. London: William Tegg, 1842); the edition is the first that would have been available to readers during the period considered in the present work.

10   Joyce (1805), 45. Joyce tutored the third Earl Stanhope's sons, Charles and James, for several years. The Earl was particularly interested in electricity and printed his own *Principles of Electricity* (1779). Joyce does not mention his own imprisonment for political activities in the 1790s.

11   Bernhard Kuhn, *Autobiography and Natural Science in the Age of Romanticism: Rousseau, Goethe, Thoreau* (Farnham: Ashgate, 2009), 62.

sales to the middle classes'.[12] John Issitt adds, too, that the price of the *Dialogues* gradually fell to a price of two shillings and sixpence for the whole set, making it available to 'a much wider audience'.[13] The changes in readership echoed those of contemporary society and print, demonstrating the interconnected nature of social, economic and publication cultures. The process of price reduction made texts such as Joyce's more popular, if not by original intention, in the sense that they were purchased by more readers. However, to describe the text as 'popular' does not convey fully or accurately the multiple processes of popularisation, for it implies a commonality between published volumes, journals, lectures and illustrations, based largely on assumptions about readers' limited education levels or desire for information. Approaching popular science in this way ignores important distinctions between the diversity of popular publication available to contemporary readers. Instead, many different approaches existed in popular texts, both in terms of writing and reading.

The rapidly growing market for published works in the 1840s produced a very different type of scientific authorship, one that sought to address complex subjects such as electricity and to convey, beyond formal institutional frameworks, developments in science to readers. Authors' scientific expertise varied considerably but they attempted, nonetheless, to write about science in ways that would be simultaneously engaging, authoritative and understandable. They also needed to satisfy the demands of newly literate readers, who were often relatively uninformed about electricity or, worse, badly misinformed. Neither were authors' motives solely altruistic, educational or performative. One of the underlying purposes in producing interesting writings about science was to generate and sustain sales. It is not possible to gauge accurately or assert the impact on book sales of the burgeoning periodicals market, due to the effect of the numerous other concurrent factors discussed previously. We can assert more confidently, though, that the massive growth of public interest in science during the period created the type of expanded demand for writings about science in both books and periodicals described by Aileen Fyfe and Bernard Lightman in *Science in the Marketplace*, which in turn stimulated more varied scientific authorship and increased quantities of information for wider and more diverse readerships.[14]

Readers were not incidental to the processes of popularisation; they were essential and active participants. Recognising their role in 'the processes by which the authoritative audience relations of science were actually accomplished' is

---

[12]    Simon Eliot, 'Some Trends in British Book Production 1800–1919', in *Literature in the Marketplace: Nineteenth-Century British Publishing and Reading Practices*, ed. John O. Jordan and Robert L. Patten (Cambridge: Cambridge University Press, 2003), 39.

[13]    John Issitt, *Jeremiah Joyce: Radical, Dissenter and Writer* (Aldershot: Ashgate, 2006), 122.

[14]    Aileen Fyfe and Bernard Lightman, *Science in the Marketplace: Nineteenth-Century Sites and Experiences* (Chicago: University of Chicago Press, 2007).

vital, as Jonathan Topham suggests, in understanding the processes themselves.[15] Interactions between authors, knowledge and readers did not comply with diffusionist and passive models of 'reception', 'diffusion' or 'transmission', concepts Stephen Shapin also rejects in favour of 'popularisation'.[16] Regardless of the terms we use, the process of popularisation was neither 'the transmission of a simplified version of research findings to a passive public' nor the 'passive lay consumption of learned products'.[17] Books about electricity were popular because, compared with their predecessors, they were affordable and they offered access to information, rather than just simplification.[18]

The presentation of scientific knowledge shaped what was available and, for that reason, understandings of electricity were not separate from how it was written about. As James Secord suggests, 'the debate about [what] forms knowledge should take was at every point implicated in the making of knowledge'.[19] The making of knowledge through experimentation, as the titles of contemporary publications suggest, contributed significantly to the dual perception of electricity as 'curious', 'singular' and 'interesting', yet also an activity that could be pursued by anyone.[20] From the 1830s, the proliferation of books on electricity prompted a contemporary commentator to declare, 'on no subject, perhaps, do text-books go sooner out of date than on the widely interesting one of electricity and its kindred phenomena of magnetism, so rapid are the strides of progress in these sciences'.[21] Popular pamphlets on electricity drew equally on lectures, experiments and

---

[15]    Jonathan Topham, 'Scientific Publishing and the Reading of Science in Early Nineteenth-Century Britain: An Historical Survey and Guide to Sources', *Studies in History and Philosophy of Science*, 31:4 (2000), 560.

[16]    Stephen Shapin, 'Social Uses of Science', in *The Ferment of Knowledge: Studies in the Historiography of Eighteenth-Century Science*, ed. G.S. Rousseau and Roy Porter (Cambridge: Cambridge University Press, 1980), 95n.

[17]    Peter Bowler, Presidential Address, 'Experts and Publishers: Writing Popular Science in Early Twentieth-Century Britain, Writing Popular History of Science Now', *British Journal for the History of Science*, 39:2 (June 2006), 164; Roger Cooter and Simon Pumfrey, 'Separate Spheres and Public Places: Reflections on the History of Science Popularisation and Science in Popular Culture', *History of Science*, 32 (September 1994), 254. See also Whitley (1985), 3–28.

[18]    For further discussion, see also Russell Berman, 'Popular Culture and Populist Culture', *Telos*, 82 (1991), 59–70; William Beik and Gerald Strauss, 'The Dilemma of Popular History', *Past and Present*, 141 (1993), 207–15; and Morag Shiach, *Discourse on Popular Culture: Class, Gender and History in Cultural Analysis, 1730 to the Present* (Stanford: Stanford University Press, 1989).

[19]    Secord (1987), 444.

[20]    William Sturgeon, *Recent Experimental Researches in Electro-Magnetism, and Galvanism: Comprising an Extensive Series of Curious Experiments, and their Singular and Interesting Results; Showing that Electro-Magnetic Action may be Developed and Modified by Processes not Genrally [sic] Known.--With Some Practical and Theoretical Observations on that Department of Science* (London: Sherwood, Gilbert and Piper, 1830).

[21]    'Electricity', *Chambers's Journal*, 123 (10 May 1856), 303.

periodical contributions, and presented electricity as a subject both practical and theoretical.[22] Explication was key, rather than original discovery; as Aileen Fyfe contends, 'in the 1830s scientific reputation was not yet so tightly tied to the publication of original research'.[23] In writing about electricity, the affiliations between experiment, research and publication were as close as those between authorship, science and reading.

The ways authors' and publishers' new awareness of readerships altered publication content and presentation is illustrated by one of the earliest books on electricity, *A Manual of Electrodynamics* (1827), a translated treatise by French mathematician Jean Firmin Demonferrand (1795–1844).[24] The translation of the French *manuel* as 'manual' usefully defined the approachability of writings about electricity beyond scientific treatises or discourses.[25] The term had deeper significance, too, in establishing associations between electrical science and practical, physical endeavours, based on first principles, rather than loftier mental or scholarly pursuits. The shift of emphasis gave a particular direction and purpose to writings about electricity, one that was cognisant of readers' circumstances and aspirations, as well as its future applicability as a science. Other early volumes as William Leithead's *Electricity: Its Nature, Operation and Importance in the Phenomena of the Universe* (1837), endowed electricity with simultaneously inspiring and useful qualities, yet protests that it also 'has no pretensions to the title of a scientific treatise'.[26] Leithead's style aimed to engage a broad readership but, while reviewers appreciated his 'plain and intelligible manner', they commented that he gets 'carried away by his enthusiasm and writes in a somewhat more elevated style ... intended, apparently, for non-scientific readers'.[27] Leithead's high rhetorical style reflected that of his lecturing, a common source for

---

[22]    For example, William Sturgeon (1830); William Ritchie, *Experimental Researches in Voltaic Electricity and Electro-Magnetism* (London: Richard Taylor, 1832).

[23]    Aileen Fyfe, 'Conscientious Workmen or Booksellers' Hacks? The Professional Identities of Science Writers in the Mid-Nineteenth Century', *Isis*, 96:2 (June 2005), 192.

[24]    Jean Firmin Demonferrand, *Manuel d'Électricité Dynamique* (Paris: Bachelier, 1823); *A Manual of Electrodynamics* (Cambridge: J.J. Deighton, 1827). The translator James Cumming (1777–1861) taught Chemistry at Cambridge from 1815 to 1860; see Mary D. Archer and Christopher D. Haley, *The 1702 Chair of Chemistry at Cambridge: Transformation and Change* (Cambridge: Cambridge University Press, 2005), 159.

[25]    See, for example, Benjamin Wilson, *A Short View of Electricity* (London, 1780); Joseph Priestley, *A Familiar Introduction to the Study of Electricity* (London: J. Johnson, 1786); Thomas Thomson and William Blackwood, *An Outline of the Sciences of Heat and Electricity* (London: Baldwin and Cradock; William Blackwood, Edinburgh, 1830).

[26]    William Leithead, *Electricity: Its Nature, Operation and Importance in the Phenomena of the Universe* (London: Longman, Orme, Brown, Green, and Longmans, 1837).

[27]    Review: 'Electricity; its Nature, Operation, and Importance in the Phenomena of the Universe by William Leithead', *British Magazine*, 12 (December 1837), 675. The reviewer's use of the term 'enthusiasm' may refer to associations with evangelical fervour.

nineteenth-century publications about electricity. Indeed, to understand the role of popular writings, we need to keep in mind the influence of these more ephemeral and oral forms, and the way in which they shaped written responses.

William Whewell, founder of the British Association for the Advancement of Science, who had recommended the translation of Demonferrand's work, provided especially authoritative interventions. According to Simon Schaffer, Whewell was 'peculiar' as a scholar because he was simultaneously literary, scientific, philosophical and critical.[28] He chose not to pursue 'a life professionally devoted to the science'; nonetheless, the 'massive and erudite' *History of the Inductive Sciences* (1837) and his subsequent volume, *Philosophy of Inductive Sciences* (1840) provided essential contributions to the experimental investigation of electricity.[29] His focus was not just on the stages of experimentation but also on how participants in scientific investigations thought, what fuelled the process of discovery, and the different ways in which scientific evidence could be interpreted. He related the progress and formulation of scientific laws to the advances of contemporary culture, presenting new discoveries as the pinnacle of intellectual sophistication and advancement, rather than as threats to the status quo.

Whewell argued that, rather than phenomena being subject to unreliable explanations based on individual conclusions, they were best pursued as a collective endeavour. His writings on electricity offered clear expositions of concepts, independent of advanced theories, for readers from a variety of educational and social origins. He argued that facts and observations were secondary to the essential connections identified between them. While the 'facts' of electricity had always been observed, in lightning, sparks or static, its true meaning and implications could only be understood if it was recognised that light, heat and magnetism stemmed from the same fundamental energy source. Whewell had 'the vague and obscure persuasion that there *must* be *some* connection between electricity and magnetism, so long an idle and barren conjecture, was unfolded into a complete theory' (Whewell's emphases).[30] For him, understanding electricity exemplified the essential relationship between the active mind and scientific progress. Scientific discovery was creative in the sense that, without the imaginative and narrative framework of the mind, there would be only a collection of unconnected facts, phenomena and observations.

---

The relevance of the association would depend on whether Leithead was a dissenter; however, this is not clear from the minimal records that exist of Leithead's life.

[28]    Simon Schaffer, 'The History and Geography of the Intellectual World: Whewell's Politics of Language', in *William Whewell: A Composite Portrait*, ed. Menachem Fisch and Simon Schaffer (Oxford: Clarendon, 1991), 230.

[29]    William Whewell, *Philosophy of the Inductive Sciences*, vol. 1 ([1840] London: Routledge/Thoemmes Press, 1996), xii; John R.R. Christie, 'The Development of the Historiography of Science', in *Companion to the History of Modern Science*, ed. R.C. Olby, G.N. Cantor, J.R.R. Christie and M.J.S. Hodge (London: Routledge, 1990), 13.

[30]    Whewell (1840), 361.

Faraday, 'the father of electricity', also contributed significantly to the landscape of books on electricity.[31] Before 1836, his papers were available only to the relatively limited audiences who could either attend his lectures or afford to purchase the Royal Society's *Philosophical Transactions*.[32] The publication of his research papers in volumes between 1839 and 1856 offered readers a vast source of expert knowledge about electricity.[33] They also demonstrated Barri Gold's point that 'many scientists had what we now call literary aspirations and acted as their own popularizers'.[34] The multiple volumes support Peter Bowler's view that the professionalisation of science and popular writing were not necessarily opposed; indeed, practitioners were able to have 'limited participation in projects which were seen to have educational merit or publicity value for science as a whole'.[35] As David Knight suggests, too, popularising only came to be somewhat 'despised' in the early twentieth century.[36] The intended audience for Faraday's published papers is not specified but he does state in the preface that they were offered 'at a moderate price … to those who may desire to have them'.[37] The egalitarian tone of Faraday's comment suggests a desire for widened scientific participation, so that others might follow his lead as a self-taught philosopher, scientist and discoverer.

Over the next 50 years, Faraday's wish was realised. A collection of writers stepped forward to popularise knowledge about electricity, through lecturing, books, pamphlets and other writings. These included Henry Minchin Noad, FRS (1815–1877), who lectured on chemistry and electricity at literary and scientific institutions in Bath and Bristol from 1836 and produced such works as *A Course of Eight Lectures on Electricity, Galvanism, Magnetism, and Electro-Magnetism* (1839), *Manual of Electricity* (1857) and *The Student's Textbook of Electricity* (1867).[38] Noad's final volume sought to correct the flawed understandings of electricity being purveyed by less-informed writers such as Robert M. Ferguson,

---

[31]    Frank Ashall, 'The Father of Electricity', in *Remarkable Discoveries!* (Cambridge: Cambridge University Press, 1994), 1–16.

[32]    Michael Faraday's first published paper on electricity was 'Experimental Researches in Electricity', *Philosophical Transactions of the Royal Society of London*, 122 (1832), 125–62.

[33]    Michael Faraday, *Experimental Researches in Electricity*, series 1–14 [*Phil. Trans.*, 1831–1838] (London: Bernard Quaritch, 1839). The volumes consisted of 30 papers Faraday wrote between 1831 and 1856. It should not be confused with William Sturgeon's *Experimental Researches* (1830).

[34]    Barri J. Gold, *ThermoPoetics: Energy in Victorian Literature and Science* (Cambridge, MA: MIT Press, 2010), 27.

[35]    Bowler (2006), 160.

[36]    Knight (2003), 75.

[37]    Faraday (1832), 2.

[38]    G.C. Boase, 'Noad, Henry Minchin (1815–1877)', rev. Iwan Rhys Morus, *Oxford Dictionary of National Biography* (Oxford University Press, 2004) [http://ezproxy.ouls. ox.ac.uk:2117/view/article/20214; accessed June 1, 2011].

author of *Electricity* (1867).[39] Noad limited the extent to which he tried to explain the actual phenomenon of electricity, claiming that the volume did not 'pretend to a scientific character, or to convey original information' but catered, instead, to 'the taste of the public in general'.[40] Doubts about the existence and nature of electricity were still evident in the 1830s, as indicated by Noad's assurance that its 'identity' was now 'decided, as to admit of no doubt in the minds even of the most sceptical'.[41]

## Rhetorical Non-Fiction

Robert Hunt's writings drew on the lectures he had given on mechanical and physical sciences, but explored other presentations of electricity as well.[42] He was well-known to contemporary readers through such volumes as *Researches on Light* (1844), *Economic Geology* (1846) and *The Poetry of Science* (1848), as well as his regular periodical contributions.[43] Hunt's *Poetry of Science* approaches electrical science and its associated technologies with unerringly optimism and, significantly, his title refers to the 'poetry *of* science', rather than 'poetry *and* science'. In the opening, he declares that the superior combination of science, poetry and philosophy goes beyond 'mere economic applications' and has, instead, the 'power of exalting the mind to the contemplation of the Universe's profound powers'.[44] In his view, man was fundamentally altered by the combination of science and literature, and he portrays this as an almost chemical or mathematical process:

> Poetry seizes the facts of the one, and the theories of the other; unites them by a pleasing thought, which appeals for truth to the most unthinking soul, and leads the reflective intellect to higher and higher exercises; it connects common phenomena with exalted ideas.[45]

---

[39]  Robert M. Ferguson, *Electricity* (Edinburgh: William and Robert Chambers, 1867).

[40]  Henry Minchin Noad, *A Course of Eight Lectures on Electricity, Galvanism, Magnetism, and Electro-Magnetism* (London: Scott, Webster and Geary, 1839), ii.

[41]  Noad (1839), iii.

[42]  Hunt rose to prominence as the Chair of experimental physics at the Royal School of Mines and became a fellow of the Royal Society in 1854. Alan Pearson, *Robert Hunt* (St Austell: Federation of Old Cornwall Societies, 1976) and Lambert M. Surhone, Mariam T. Tennoe and Susan F. Henssonow, eds, *Robert Hunt (Scientist)* (Saarbrücken: VDM Verlag Dr. Mueller AG and Co. Kg, 2010).

[43]  Robert Hunt, *The Poetry of Science, or Studies of the Physical Phenomena of Nature* (London: Reeve, Benham and Reeve, 1848); Hunt regularly contributed to the *Art Journal*, *The Athenaeum*, the British Association's reports, the *Pharmaceutical Times*, *Sharpe's London Magazine of Entertainment and Instruction* and the *Photographic Journal*.

[44]  Hunt (1848), vii.

[45]  Hunt (1848), xviii.

Hunt's text is a performance, a type of contemporary writing James Secord describes as the 'rhetoric of spectacular display', but one that also met with criticism.[46] A reviewer confessed in 1849, for example, that he had difficulty seeing 'the realities of matter with the same enthusiastic eye' as Hunt, and that he doubted Hunt's conclusion that 'every scientific truth is essentially poetical'.[47] The reviewer concludes that Hunt, whom he professes to know little about, was more a poet than a philosopher and that his partiality must be due to limited technical understanding. In fact, as Hunt's publications indicate, he was adept in a number of different emerging sciences, often at relatively advanced levels.

Such hyperbole as Hunt's was not the only obstacle apparent in the transitional morass that lay between scientific and literary works. The *Athenaeum* reported in 1850, for example, that 'the very discursive system' of the mechanics' institutes 'has been fraught with evil consequences', such as 'a superficial acquaintance with the pursuits of science' and 'a lamentable and vicious dilettantism'.[48] Works such as Robert Weale's series, *Rudimentary Works for Beginners* (1848), sought to address the gaps in public knowledge left by public lectures and other less reliable forms of information.[49] The volume on electricity in the series, Sir William Snow Harris's *Rudimentary Electricity* (1850), was deemed to provide 'a sound and practical view of the subject' and 'essentially a contradiction to the pernicious doctrine of its danger'.[50] The existence of a market for yet more explanations of electricity indicates the persistent tensions that existed, not just around the correctness of scientific theories and the threat to scientific integrity of popular versions and speculations, but also in relation to the potentially disturbing and even supernatural qualities that electricity might possess. The difficulties had by contemporary readers in negotiating electricity's properties are evident in the *Edinburgh Review*'s review of Harris's work.[51] While readers' interest in technological applications is foregrounded, in electricity being described as the

---

[46]    James A. Secord, *Victorian Sensation: The Extraordinary Publication, Reception, and Secret Authorship of Vestiges of the Natural History of Creation* (Chicago: University of Chicago Press, 2000), 439, 98.

[47]    Review: 'ART. IV.-1. The Poetry of Science; or, Studies of the Physical Phenomena of Nature', *North British Review*, 13:25 (May 1850), 120.

[48]    Review: 'Rudimentary Works for Beginners', *Athenaeum*, 1196 (28 September 1850), 1025.

[49]    Other volumes in John Weale's series were *Divers Works of Early Masters in Christian Decoration* (1846) and *Rudimentary Dictionary of Terms used in Architecture, Building, and Engineering* (1849–1850).

[50]    Review (1850), 1025; William Snow Harris, Sir (1791–1867), *Rudimentary Electricity, being a Concise Exposition of the General Principles of Electrical Science* (London: John Weale, 1848). Harris's volume was revised and enlarged in 1872 by Henry Noad.

[51]    Review: Art. IV.-1. 'Rudimentary Electricity; being a Concise Exposition of the General Principles of Electrical Science, and the Purposes to which it has been Applied', *Edinburgh Review*, 90:182 (October 1849), 441.

'letter-carrier and message-boy' of the age, we are also told that magnetism 'is unclothed of mystery, and set to drive turning lathes'. The reviewer asks the persistent question of 'what is electricity?' but also comments that 'the distracted reader, who finds one electricity perplexing enough, loses count and heart, and closes the treatise in despair'. [52] Learning about electricity was confusing enough without having to deal with a multiplicity of conflicting accounts that may or may not be true, new concepts and new terms. As Faraday, Maxwell and Whewell had recognised previously, emerging ideas and terminologies in electrical science could represent serious deterrents to learning about electricity, lending even more instability to a phenomenon that was already prone to endemic speculation.

In the unstable understanding about electricity, treatises and books continued to be important sources of information about electrical science. The popularisation of *mis*understandings about electricity – rather than understandings – represented exactly the type of 'intellectual anarchy' John Stuart Mill had warned of at the beginning of the decade.[53] While changes in publication climates reduced the cost of book production and enabled the expansion of publication forums beyond the book, price was only one aspect of readers' access. New publications forms both catered to and reflected the changes of participation in electrical investigations and science. In the 1870s, Faraday's former protégé John Tyndall published two popular volumes on electricity.[54] His discussions also moved away from questions of electricity's operations, and towards the newly invented electrical technologies and machines.[55] The new alliances between science and other cultural authorities such as economics, manufacturing and politics meant that the scientist's role was increasingly understood as a mediation between science and its applications in technological developments.

Popular books frequently offered expert knowledge about electricity, supporting Cooter and Pumfrey's proposition that popularisation did not convey simply 'watered down' knowledge.[56] However, the two types of knowledge are also not as distinct as they might at first appear; even today, assumptions about scientific knowledge continue to plague understandings of how, why and by whom science is produced and popularised. As Richard Whitley observes, 'essentially, popularisation is not viewed as part of the knowledge production and validation process but as something external to research which can be left to non-scientists,

---

[52] Review (1849), 445.

[53] J.S. Mill, 'The Spirit of the Age', *Examiner*, 3:2 (13 March 1831), 162–3.

[54] *Notes of a Course of Seven Lectures on Electrical Phenomena and Theories* (London, 1870); *Lessons in Electricity at the Royal Institution, 1875–6* (London: Spottiswoode and Co., 1876).

[55] Tyndall discusses three areas of advancement: firstly, technologies, such as the telegraph, electro-chemistry and electrolysis; secondly, electrical theories, such as Ohm's Theory, Faraday's Electrolytic Law and Nobili's Iris Rings; and, thirdly, electrical machines, such as the condenser, magneto-electric machines and the Leyden battery.

[56] See Cooter and Pumfrey (1994).

failed scientists or ex-scientists as part of the general public relations effort of the research enterprise'.[57] Whether the information they offered was correct or not, popular books sustained nineteenth-century interest in electricity and in science more generally. Their work also represented an emergent discourse that created vital connections between the period's scientific exploration and the development of its literary and publication forums.

Non-fiction is often excluded from studies of 'literature' and 'literary' writing, restricting the usage of the two terms, as Ralph O'Connor notes, to a relatively limited range of fictional works, such as novels, poems and plays.[58] The non-fiction responses to electricity discussed in this chapter illustrate literary qualities, not just in writings about science but also in the reading methods upon which they relied and the interrelated changes in social and publication contexts, particularly the emergence of the popular periodical. The variety of periodical publications, contributors and readers between the 1830s and 1880s negates any possibility of a single, unified or representative stance. I aim to convey, instead, the varying approaches to the subject and the degrees of knowledge apparent in periodical writings, by examining relevant articles from four markedly different publications: the *Annals of Electricity*, *Reynolds's Miscellany*, *Punch* and *Once a Week*. I also investigate the extent to which non-specialist writers relied on scientific understandings of electricity and how the context of electrical science in the periodical allowed authors to explore new avenues of engagement with the phenomenon.

The emergence of the periodical press is inseparable from the wider context of contemporary publishing, which it is important to take into account. The British periodical press was a site of significant change, particularly in the mid-nineteenth century, when its dramatic expansion was prompted by the combined force of new reading audiences, innovations in printing and transport technologies, and reduced publishing costs; as Dawson, Noakes and Topham point out, newspaper taxes were reduced in 1836 and repealed in 1853 and 1855, while taxes on paper and rags were abolished in 1860 and 1861.[59] The increase in cheap, frequent publications was both produced for and driven by the emergence of a more broadly literate mass audience. The periodical is now recognised as a crucial element of contemporary responses to innovations such as electricity. Their significance and cultural location is increasingly understood, in that 'readers outside the relatively small and elite intellectual community depended largely on magazines, periodicals, and newspapers for their understanding of contemporary

---

[57]  Whitley (1985), 3.

[58]  Ralph O'Connor, *The Earth on Show: Fossils and the Poetics of Popular Science, 1802–1856* (Chicago: University of Chicago Press, 2007), 14.

[59]  Gowan Dawson, Richard Noakes and Jonathan R. Topham, 'Introduction', in *Science in the Nineteenth-Century Periodical: Reading the Magazine of Nature*, ed. Geoffrey Cantor, Gowan Dawson, Graeme Gooday, Richard Noakes, Sally Shuttleworth and Jonathan R. Topham (Cambridge: Cambridge University Press, 2004), 16.

cultural issues'.[60] Secondly, there is greater recognition that scientific topics, to which electricity often related, were widely discussed beyond specialist forums. As scholars have noted, 'general periodicals probably played a far greater role than books in shaping the public understanding of new scientific discoveries, theories, and practices'.[61] Nonetheless, the heterogeneity of periodical contents continues to elude easy categorisation and makes analysing them uniquely challenging. As Lyn Pykett notes, periodical scholars require a seemingly impossible 'total knowledge of a past culture' and 'a degree of conceptual possession of a "documentary" culture which must elude them even in relation to a living culture'.[62] Today's digitisation and new technologies allow considerable headway to be made in studying the nineteenth-century periodical though, as I now discuss, considerable methodological obstacles still remain.

One method of ascertaining the prominence of electricity in periodicals is to identify and survey when, where and how writers refer to the phenomenon. Before examining individual writings, therefore, we can note the extent to which electricity was mentioned in both non-fiction and fiction articles of the period.[63]

Table 3.1 Example keyword search: *Periodicals Archive Online* (citations and article full text; January 1831 to December 1881)

| Keyword | Articles | Book reviews | Other (unindexed front and back matter) |
|---|---|---|---|
| Electricity | 11,917 | 138 | 3,259 |
| Electrical | 6,556 | 68 | 1,189 |
| Electric | 20,632 | 195 | 3,322 |
| **Total** | **39,105** | **401** | **7,770** |

---

[60] Geoffrey Cantor, Gowan Dawson, Richard Noakes, Sally Shuttleworth and Jonathan R. Topham, 'Introduction', in *Culture and Science in the Nineteenth-Century Media*, ed. Louise Henson, Geoffrey Cantor, Gowan Dawson, Richard Noakes, Sally Shuttleworth and Jonathan R. Topham (Aldershot: Ashgate, 2004), xvii.

[61] Dawson et al. (2004), 1–2.

[62] Lyn Pykett, 'Reading the Periodical Press: Text and Context', in *Investigating Victorian Journalism*, ed. Laurel Brake, Aled Jones and Lionel Madden (Basingstoke: Macmillan, 1990), 5.

[63] *British Periodicals Online* and the *Periodical Archive Online* are the most relevant sites in this instance, due to the range of publications to which they refer and their use of digitised originals. Other online archive resources are valuable in different ways: *Science in the Nineteenth-Century Periodical* is more finely tuned and highly effective for investigating the site's selection of publications, by means of modern summaries rather than the original texts, while *19th-Century UK Periodicals* is well-suited to broader explorations that do not require specifying the genre of the work.

Table 3.2 Example keyword search: *British Periodicals Online* (citations, excerpts and article full text; January 1831 to December 1881)

| Keyword | Advertisements, articles, graphics, reviews | Drama, fiction, poems |
|---|---|---|
| Electricity | 9,850 | 542 |
| Electrical | 5,161 | 408 |
| Electric | 14,363 | 2,180 |
| **Total** | **29,374** | **3,130** |

Identifying references by means of relevant keywords provides a degree of unrefined information, based on the limited generic definitions of computerised search facilities. It makes it possible to deduce that the term 'electric' may have been used more frequently than that of 'electrical' (see Tables 3.1 and 3.2), or that electricity was referred to more frequently in non-fiction than fiction (see Table 3.2). A degree of further analysis is possible in terms of trends, genres and individual publications. However, identifying periodical references in this way reveals at least two methodological problems. Firstly, the volume of references to electricity makes it difficult to assess how representative selected examples are, and, secondly, for the significance of references to be properly evaluated, they would need to be categorised more tellingly – a process beyond the purposes of the present work. Thirdly, as previously noted, writings about electricity have a tendency to defy this type of specific classification. Keyword searches are a useful means of initial identification but ascertaining which references offer real insights into contemporary perceptions of electricity involves more extensive scrutiny, comparison and analysis. An examination of several thousand individual instances indicates that many references to electricity were peripheral to the main content of the article, and that terms relating to electricity were frequently used in an unspecific or metaphorical sense, or with reference to topics on the margins of my scope here.[64] A great many other references simply repeat the history of discoveries about electricity, with little further reflection or progression.

The publications discussed in this chapter have been chosen because they engage at length and in depth with the subject of electricity. The aim is to illustrate how non-fiction writers used the literary possibilities of the periodical to shape popular awareness of ideas about electricity, through editing and formatting, serialisation and illustrations, and to consider what insights they offer into contemporary perceptions of electricity and experimentation. The place held by science in nineteenth-century periodicals was particularly appropriate for writings about electricity. As Richard Yeo suggests, not only were periodicals 'an

---

[64]    Topics discussed include the telegraph, light, atmospheric electricity in lightning or fog, spiritualism and 'planchette', now more commonly known as a 'Ouija' board.

indispensable forum for science' but 'outside the major periodicals there were even more divergent notions of science'.[65] Continuing speculation about electricity and uncertainties about its scientific basis allowed it to be viewed from a variety of perspectives and to find a venue in the increasing wide-ranging periodicals market. Writings about electricity revealed an assortment of social and political tensions surrounding its perception, investigation and representation.

The authorship of periodical contributions, as Matthew Rubery points out, was never 'straightforward'; more often it involved 'collaboration among proprietors, editors, writers, illustrators, and even advertisers'.[66] This diversity complicated the specialisation of the electrical sciences, as did the emergence of electricians, a group one contemporary medical journalist describes as those 'educated to the lower grades of the profession'.[67] The new practitioners were perceived as a new breed of investigator, for their interest was electrical measurements and technological applications, rather than the electricity's fundamental properties. From the 1830s onwards, writers discussing electricity in periodicals bolstered their scientific and literary authority by referring to public demonstrations and well-known figures at the Royal Institution or the London Electrical Society. As James Secord notes, 'it is increasingly evident that the usual categories used to define scientific writing, such as "popular science", "specialist science", "original discovery", are inadequate'.[68] Like the authors of books on electricity, popular writers did not report the spectacle of scientific experimentation merely for readers' entertainment or edification; many also engaged fully with the specialist lectures and papers they described, further blurring the distinctions between popular and specialist science.

## Electrical Communities

Periodicals acted as an important resource for those who wanted to learn, read and write about electricity. The monthly *Annals of Electricity, Magnetism and Chemistry, and Guardian of Experimental Science*, for example, was published between 1836 and 1843 in manageable 80-page issues, priced at around two shillings, and edited by the formidable William Sturgeon.[69] Despite lasting just

---

[65]  Yeo (1993), 46.

[66]  Matthew Rubery, 'Victorian Print Culture, Journalism and the Novel', *Literature Compass*, 7 (2010), 292.

[67]  Untitled, *The Lancet*, 362 (16 December 1826), 2; Untitled, *London, Edinb. and Dublin Philos. Mag.*, 4th Ser., 33 (1867), 397 (OED).

[68]  James Secord, ed., *Collected Works of Mary Somerville*, vol. 1 (Bristol: Thoemmes Continuum, 2004), xxvi.

[69]  The Waterloo Directory entry states the price as one shilling; however, the prices stated on the original editions are between two shillings two pence and, on later issues, two shillings sixpence.

seven years, it is notable as the only nineteenth-century periodical specifically devoted to electricity and because its contributors were experimentalists who constituted something of an electrical community. William Sturgeon's editorship also represents an intriguingly literary-scientific performance, which subverts conventional divisions between writing, editorship and reading.

Scientific integrity was paramount, so much so that the title of the *Annals* declared it to be the 'guardian of experimental science'. Experimental proofs were central to the journal's mission of vanquishing superstition and the supernatural from investigations of electricity. As Geoffrey Cantor suggests, 'the appeal to experiment is generally the most effective persuasive strategy when arguing for or against some theory or doctrine' and 'the rhetoric of experiment is endowed with considerable power'.[70] Nonetheless, as Sturgeon warned his contributors and readers, they should avoid 'whimsical hypotheses which have no reality in nature', and his advice indicates the menaces that speculation and imagination were perceived to be in studying electricity.[71]

As a subject of scientific study, electricity was a relative newcomer, in an environment that was more competitive and quarrelsome than we might now realise. Sturgeon criticised nineteenth-century institutions as 'very far from being well-adapted' to experimentation in specialist sciences like electricity and he asked what Joseph Priestley, 'the first electrician of the age', would have thought of electricity having becoming 'a parent of other sciences' yet one still not 'deemed worthy of a separate establishment among their temples'.[72] The ownership of electrical experimentation had yet to be securely claimed and the real enemy of progress in electrical science was, Sturgeon felt, restricted access to knowledge. He dismissed as 'groundless' and 'detrimental' the view that only those 'who are deeply skilled in experimental investigation can be really useful to science', and argued that such prejudices were especially inappropriate in experimental electrical sciences.[73] The investigation of electricity was clearly contentious, as Willis indicates, in terms of its connections with class, education and power.[74] Sturgeon's colleague Andrew Crosse also wondered 'what rank in the tree of science electricity is to hold' and declared that 'electricity is no longer the paltry, confined science which

---

[70]    Geoffrey Cantor, 'The Rhetoric of Experiment', in *The Uses of Experiment: Studies in the Natural Sciences*, ed. Trevor Gooding, Trevor Pinch and Simon Schaffer (Cambridge: Cambridge University Press, 1989), 161.

[71]    William Sturgeon, 'Address to the General Meeting of the London Electrical Society', *AoE* 2:11 (7 October 1837), 64–72. Sturgeon refers to Joseph Priestley's *The History and Present State of Electricity* (London, 1767).

[72]    Sturgeon (1837), 66–7.

[73]    William Sturgeon, Review: 'Electricity; its Nature, Operation, and Importance in the Phenomena of the Universe by William Leithead', *Annals of Electricity*, 2:11 (October 1837), 71.

[74]    Martin Willis, *Mesmerists, Monsters, and Machines: Science Fiction and the Cultures of Science in the Nineteenth Century* (Kent, OH: Kent State University Press, 2006), 71.

it was once fancied to be'.[75] Future understandings of electricity could only be achieved in parallel with social and educational change; as David Gooding asserts, 'empirical access to Nature is both a cognitive and social process'.[76] The practice of electrical investigation was all the more controversial because it involved increased participation, and that effectively destabilised restricted and privileged access to information.

Disputes were not only about class and hierarchy; debates about the different approaches of scientists and of practitioners were also rife in the history of electrical science. The several disagreements between Sturgeon and Faraday about experimental practice and methods of interpretation, conducted in the pages of the *Annals*, have been documented by Iwan Rhys Morus.[77] Even the discovery of electromagnetism and the invention of the electromagnetic motor were contested and are still sometimes credited to Sturgeon rather than Faraday.[78] Neither were Sturgeon and Faraday the only contenders for the discovery of electromagnetism; as Frank James notes, when Faraday publicised his discovery of the principle in the *Quarterly Journal of Science*, 'a rumour went around suggesting that he had plagiarized some of Wollaston's electro-magnetic work'.[79] The desire to define electricity's substance and to take ownership of explanations made electrical investigation a site of fierce competition, with individual claims of discovery often determined by highly subjective factors, such as popularity, patronage and publication.

Sturgeon, as James Secord notes, 'lambasted the formality of the existing periodicals'.[80] The *Annals* provided, in contrast, an exciting and accessible mix of scientific papers, explanations of experiments, correspondence, tables of experimental results, translated research papers and practitioners' notes. Rather than the *Annals*' form lacking a 'narrative component', it allowed more innovative presentations of electrical science and entirely new relationships with readers.[81] However, small-scale electrical experiments are frequently connected in the journal to wider topical interests, for example, in the series of experiments Sturgeon conducts from month to month on clay provided by Robert Were Fox, and the observations he relates to natural geological phenomena and industrial

---

[75]   Andrew Crosse, 'Electrical Society', *Literary Gazette*, 1097 (27 January 1838), 54.

[76]   David Gooding, '"Magnetic Curves" and the Magnetic Field: Experimentation and Representation in the History of a Theory', in Gooding et al. (1989), 192.

[77]   Iwan Rhys Morus, *Frankenstein's Children: Electricity, Exhibition, and Experiment in Early-Nineteenth-Century London* (Princeton: Princeton University Press, 1998), 43–69.

[78]   Lance Day and Ian McNeil, eds, *Biographical Dictionary of the History of Technology* (London: Routledge, 2003), 1179.

[79]   Frank A.J.L. James, *Michael Faraday: A Very Short Introduction* (Oxford: Oxford University Press, 2011), 39. Neither Wollaston nor Sturgeon appears to have openly contended Faraday's claim.

[80]   Secord (1987), 455.

[81]   Hayden V. White, *The Content of Form: Narrative Discourse and Historical Representation* (Baltimore: Johns Hopkins University Press, 1987), 11.

productivity.[82] The physical form of the periodical was important, too, in engaging and maintaining readers' interests. Each monthly issue was carefully indexed and, as a physically smaller and shorter publication than many, readers could scan easily back and forth between the contents and issues.[83] Experiments were effectively serialised and cross-indexed through editorial references, creating a rich and connected narrative and anticipation about projected experimental results that the format of the book could not offer. Precise dating of contributions and experiments in the *Annals* conveyed their immediacy and meant they could be related to whatever was happening in readers' own lives at the time. Meanwhile, the letters gave readers unprecedented access to the discursive familiarity between specialist scientists. Readers could share the thinking processes and informal camaraderie of leading scientific innovators, for example, Frederic Daniell's 1836 letter to Faraday where he pauses in his description of experiments on voltaic electricity, to say 'before I state the results, I wish to direct your attention to some observations', which he then goes on to explain.[84] The new accessibility dismantled many of the formal distinctions between knowledge producers and consumers, giving what Sturgeon referred to as 'every electrician' an alternative to the passive reception of expertise.[85]

The journal's open competitions sought 'to stimulate and promote experimental enquiry'.[86] In January 1841, one of the prize-winners was a very young James Prescott Joule, whose subsequent work on the conservation of energy led to the development of the first law of thermodynamics.[87] Sturgeon's insistence, however, that the study of electricity did not benefit from 'making too nice distinctions between discoveries or inventions' worked against the tide that would increasingly separate specialist approaches from the more generalist.[88] The journal failed in 1843, when Sturgeon ceased to cater to a broad readership by including excessively long

---

[82]    Robert Were Fox, 'Experiments on the Influence of Electrical Action upon Clay', *AoE*, 2:7 (January 1838), 54; William Sturgeon, 'Miscellaneous Articles', *AoE*, 2:11 (May 1838), 395; William Sturgeon, 'Lamination of Clay by Electricity', *AoE*, 3:14 (August 1838), 159, 161.

[83]    The *AoE* page size was 22cm, with an average 80 pages per issue; the Royal Society's *Philosophical Transactions*, in contrast, was 29cm with considerably more pages.

[84]    Frederic Daniell, 'Professor Daniell's Additional Observations', *AoE*, 1:18 (April 1836), 102.

[85]    Sturgeon (1832), 68.

[86]    [William Sturgeon], 'Prize Volumes of the Annals of Electricity, & c'., *AoE*, 6:31 (January 1841), 80.

[87]    James Prescott Joule, FRS (1818–1889); the other prize-winner was the little-known experimentalist William Henry Weeks/Weekes (1790–1850). I am indebted to the IET archivists, particularly Bill Burns, for their expert assistance in establishing the identity of Weeks.

[88]    [William Sturgeon], 'Award of Prizes', *AoE*, 6:36 (June 1841), 512.

and increasingly complex articles.[89] While Steven Shapin argues that 'the notion of a clear distinction between expert scientist and lay audience is inappropriate in the early Victorian period', Sturgeon learned to his cost that readers' interest was directly related to their knowledge of the subject.[90] Despite it being possible to read about electricity with relatively limited knowledge, the declining popularity of the journal indicates that many readers had neither the knowledge nor interest to keep up. Instead, a new demand had emerged, one that related less to the pursuit of advanced electrical research than to experiments readers could perform themselves, as participants in the future of technological applications.

## Narratives of Explanation

Popular writings about scientific developments were narratives that called upon readers to imagine experiments, as well as perform them. A particularly long-running series by Anthony Peck, published between 1846 and 1847 in *Reynolds's Miscellany*, was called 'Papers on Popular Science'. Like others in *The Saturday Magazine*, *The Art Journal*, *The Westminster Review* and *The Penny Magazine*, Peck's series sought not only to convey accurate scientific information but also to arouse and encourage readers' participation in its development. In Martha Turner's study of the relationship between mechanistic science and fiction narratives, she suggests that, from the late eighteenth century, writers 'saw themselves as engaged in a project consciously designed to provide an alternative to scientific methodology and a technologically oriented civilization'.[91] Yet writings on electricity indicate that 'scientific methodologies' were not necessarily opposed to desirable social developments; explanatory writings performed social and literary functions, as well as scientific ones, and established new relationships between writers and readers. Naturally, these purposes did not always work in harmony and, to some extent, established an inherently bi-cultural conflict.[92] Just as often, though, the representation of scientific developments was an integral feature of popular writings. As Greg Myers proposes, 'popularisers do not simply transmit or water down the writing of professionals; they transform scientific knowledge

---

[89]    The removal of the regular 'Elementary Lectures on Electricity' and other shorter items made room for items such as Daniell to Faraday, 'Sixth Letter on Voltaic Combinations', *AoE*, 10:57 (March 1843), 232–40 and 10:58 (April 1843), 241–53; Robert Kane, 'Contributions to the Chemical History of Palladium', communicated by Francis Bailey, *AoE*, 10:58 (April 1843), 253–71.

[90]    Steven Shapin, 'Science and the Public', in *Companion to the History of Modern Science*, ed. R.C. Olby, G.N. Cantor, J.R.R. Christie and M.J.S. Hodge (London: Routledge, 1990), 991.

[91]    Martha A. Turner, *Mechanism and the Novel: Science in the Narrative Process* (Cambridge: Cambridge University Press, 1993), 8.

[92]    Iwan Rhys Morus, 'The Two Cultures of Electricity: Between Entertainment and Edification in Victorian Science', *Science and Education*, 16 (2007), 593–602.

as they put it in new textual forms and relate it to other elements of non-scientific culture'.[93] Authors such as Peck publicised new understandings and, at the same time, appealed to wider readerships by amalgamating edification and entertainment.

Peck's series began in the first edition of *Reynolds's Miscellany*, which editor and proprietor G.W. Reynolds announced as follows:

> Stimulated by the growing improvement in the public taste, and convinced that the readers of Cheap Literature are imbued with a profound spirit of inquiry in respect to Science, Art, Manufacture, and the various matters of social or national importance, the Projector of this 'Miscellany' has determined to blend Instruction with Amusement; and to allot a fair proportion of each Number to Useful Articles, as well as to Tales and Light Reading.[94]

Reynolds's shameless flattery and sought to address a newly created niche whereby 'Cheap Publications contain too much light manner' and 'Periodicals are too heavy'.[95] As John Feather explains in his study of British book production and markets in the eighteenth and nineteenth centuries, 'new physical forms of printed matter, together with the newspapers and magazines, had a wider appeal primarily because they were cheaper' and because they 'were not designed for a cultural élite but for the contemporary equivalent of a mass market'.[96] Affordably priced at a shilling, *Reynolds's Miscellany* claimed 'to steer the medium course' and, in doing so, provide a new forum for writings about science. Reynolds's argument that the journal 'must provide a literary aliment suited to the improved taste of the present day' ushered in writings that encroached upon ostensibly less literary realms, such as science, trade and manufacturing. By 1855, his approach had been proved correct with the publication's sales reaching 30,000 issues a week.[97]

Peck's writing style differed from that of earlier writers such as Hunt or Oersted, who tended to employ elaborate figurative language in discussions of electricity.[98] In periodicals, the purpose was explanatory but the focus was on how electricity was produced, rather than its nature as a phenomenon. More than half of Anthony Peck's 'Papers on Popular Science' dealt with electricity, its applications and technological development and, while his identity is now sadly lost to the

---

[93]    Greg Myers, 'Science of Women and Children: The Dialogue of Popular Science in the Nineteenth Century', in *Nature Transfigured: Science and Literature, 1700–1900*, ed. John Christie and Sally Shuttleworth (Manchester: Manchester University Press, 1989), 171.

[94]    G.W. Reynolds, 'To Our Readers', *Reynolds's Miscellany of Romance, General Literature, Science, and Art*, 1:1 (7 November 1846), 16.

[95]    Reynolds (1846).

[96]    John Feather, 'The British Book Market, 1600–1800', in *A Companion to the History of the Book*, ed. Simon Eliot and Jonathan Rose (Malden: Wiley-Blackwell, 2009), 238.

[97]    *Waterloo Directory of English Newspapers and Periodicals, 1800–1900* [accessed 6 May 2011].

[98]    Hunt (1848) and Hans Christian Oersted, *The Soul in Nature* (1852).

mists of time, it is telling that he began the series with the question: 'what *practical* use is the science of electricity, the reader may perhaps say?' (author's emphasis).[99] The question prioritises, firstly, the application of science about electricity, rather than just its properties, and, secondly, the reader's role in the investigation, making a collaborative venture of it. Their interests are paramount, he seems to say, rather than his own, obscuring still further distinctions between the production of scientific writings and their consumption. The increased prominence of electricity's practical purposes brought forth a diversity of scientific discourse and participation; as Patrice Flichy points out, science could be written about 'not only by technicians but also by "literary persons": novelists, popularisers, journalists and so on'.[100] Like reviews, non-fiction periodical writings were not necessarily extensions of specialist science; instead, they represented new forms of discourse about electricity, which functioned somewhere between the two.

Peck's opening article offers several insights into contemporary presentations of electricity and their readerships, in clarifying that 'galvanism, electro-magnetism, &c. are simply branches of the science, and that the parent stem is – electricity'.[101] The fact that he feels the need to specify this connection over a decade since the initial discovery of electromagnetism indicates the considerable uncertainty that continued to exist about electricity and its associated sciences. Peck's confident assertion that electricity differed from galvanism and electromagnetism implies his expert authority but, in some ways, he also works against this impression. He claims, for example, that he is not writing for 'the information of those who are versed in the science', indicating just how closely the purposes of periodical readership related to writers' purposes.[102] He represents the production and consumption of writing as a shared venture and proposes that 'we shall studiously avoid entering into a disquisition on the merits of this or that theory; but restrict ourselves to a plain statement of the acknowledged facts'.[103] Peck's use of 'we' and 'ourselves' is ambiguous – which 'we' does he mean? Is he referring, on the one hand, to the 'we' who really know about science and condescend to pass on that knowledge to less-informed readers, or to the shared 'we' of himself and his readers? He appears to refer to both groups simultaneously, in a way that complicates the diffusionist model of 'popularisation', in which scientific ideas emanate from experts who discover and validate them, and then communicate them to passive

---

[99]   Anthony Peck, 'Popular Papers on Science: I. Electricity', *Reynolds's Miscellany*, 1:8 (26 December 1846), 125. It is likely that Anthony Peck also authored the volume *Mechanics* (London, 1846).

[100]   Patrice Flichy, *Understanding Innovation: A Socio-Technical Approach* (Cheltenham: Edward Elgar Publishing, 2007), 125.

[101]   Peck (1846), 125.

[102]   Peck (1846).

[103]   Peck (1846).

non-expert readers.[104] As Jonathan Topham suggests, the diffusionist model is 'utterly inadequate as a characterization of the actual processes of scientific communication'.[105] Certainly, the model does not reflect accurately the complex audience-relations between Peck and his readership. He does not propose that the 'acknowledged facts' about electricity emerge from more complex engagements with electrical theory; instead, he sets up an opposition between theoretical science and the more accessible 'plain statement'. He implies that the latter is preferable to the scientific 'disquisition' but that the difficulty of separating the two discourses is also evident. In the end, he concedes that 'we shall be unable wholly to exclude' technical terms and, while he deems this acceptable as long as 'they are uniformly used in scientific works', he warns that 'we should be careful to accompany them with the requisite explanation'.[106] As Richard Yeo notes of nineteenth-century science more generally, 'the institutional and educational status of science became more secure [but] this did not mean that the problem of authority was resolved; rather, it assumed different forms'.[107] The recurrent caveats Peck felt it necessary to include indicate the extent to which writing about science for wider audiences involved a troubled negotiation of both authority and space.

Instead of claiming that scientific investigations of electricity were interesting because they advanced science, popular discussions often focused firmly on practical applications. Peck distances his discussion from the superstition and speculation with which it might still be associated by referring to 'the science of electricity'.[108] By the 1840s, according to other writers at the time, there were 'few persons who have not seen an electrical machine and witnessed the spark which passes from it'.[109] Yet experiments on electricity, optics and magnetism were also still thought – albeit often satirically now – to have 'a kind of magical appearance', one which 'among the ignorant and credulous might easily pass for miracles'.[110] Peck emphasises that the importance of scientific investigation 'is materially increased by the utility which it is found to possess'.[111] The advent of electrometallurgy in the 1840s made the battery a type of an 'economic machine', Iwan Morus suggests, whereby electricity increasingly became 'a matter of commercial, economic speculation', especially in terms of the financial rewards

[104]    See Simon Schaffer, 'Scientific Discoveries and the End of Natural Philosophy', *Social Studies of Science*, 16 (1986), 407, and Jan Golinski, *Science as Public Culture: Chemistry and Enlightenment in Britain, 1760–1820* (Cambridge: Cambridge University Press, 1992), 9.

[105]    Topham (2000), 560.

[106]    Peck (1846), 125.

[107]    Richard Yeo, 'Science and Intellectual Authority in Mid-Nineteenth-Century Britain: Robert Chambers and "Vestiges of the Natural History of Creation"', *Victorian Studies*, 28:1 (Autumn 1984), 9, 31.

[108]    Yeo (1984).

[109]    'Morse's Magnetic Telegraph', *John Bull*, 1:230 (6 July 1844), 426.

[110]    'On the Origin of the Black Art, or Magic', *The Penny Satirist*, 375 (22 June 1844), 3.

[111]    Peck (1846), 125.

of nineteenth-century electrical exhibitions and spectacles.[112] However, it should be noted, too, that the recognition of electricity's commercial potential took place very gradually, and that the relevance of scientific developments and participation was related directly to cost. Peck and other popular writers on electricity described how to make electrical machines but noted also that it was 'so very expensive a piece of apparatus, and so liable to accident, that few of our readers will probably be induced to purchase one from a mathematical instrument maker'.[113] Electricity's promise as a 'moving power' was also related to matters of convenience and cost reduction. Peck describes, for example, how the work of German electrical experimenter Emil Stöhrer was not for some grand scientific vision but, rather, in the hope that electricity could reduce the five-pound cost of steam transportation from Leipzig to Dresden to a more affordable six shillings.[114] Such apparently peripheral aspects of electricity and experimentation reveal the distinctive engagements with consumer culture evident in popular writings about electricity.

Peck may have tried to distance his investigations and his writing from specialist accounts of electricity but he does refer to core conceptual developments by such as Benjamin Franklin, Michael Faraday, Luigi Galvani and Alessandro Volta. At times, he does not seem entirely convinced that, for example, 'modern investigations' have proved 'beyond a doubt that whatever electricity may be, it exists more or less in every state of the atmosphere, and in all known substances'.[115] His use of the phrase 'whatever electricity may be' hints at the inherent difficulty of understanding and representing the phenomenon, an ambiguity evident also in his claim that electricity 'is generally supposed to be a subtle fluid, pervading matter in all its states'.[116] In both the specialist and popular communications, conceptualising electricity was prone to many of the same obstacles; when even knowledge practitioners like Peck struggled to explain the fundamentals of electricity, it is easy to see why non-specialist readers continued to be confused. The nature of writings such as Peck's was neither fixed nor predictable; in fact, they frequently demanded to be read on two quite different levels. Some of the simpler experiments could be performed by readers as they read, much as a recipe is read while cooking or an instruction manual is followed when fixing an appliance (see Figure 3.1).

---

[112]   Morus (1998), 167, 165.

[113]   'Simple Electrical Machines', *Peter Parley's Annual: A Christmas and New Year's Present for Young People* [Date Unknown], 51 [19th-Century UK Periodicals; Gale Document Number: DX1901717113].

[114]   Emil Stöhrer (1812–1890) was active in the development of batteries, generators, optics and the telegraph; Stöhrer's many achievements are described by Anton A. Huurdeman in *The Worldwide History of Telecommunications* (London: Wiley-IEEE, 2003).

[115]   Peck (1846), 125.

[116]   Peck (1846).

### POPULAR PAPERS ON SCIENCE.

By ANTHONY PECK, B.A.

No. I.—ELECTRICITY.

OF what *practical* use is the science of electricity, the reader may perhaps say? In reply to such a question, we beg to refer him to a few inventions which have been effected by its agency. We allude to the electrotype, the electric telegraph, the application of electricity to agricultural purposes, and to its employment in some cases as a moving power. These inventions, however, it may be said, have emanated from the application of galvanic power, electro-magnetism, &c. This is very true: yet, at the same time, it must be borne in mind that galvanism, electro-magnetism, &c., are simply branches of the science, and that the parent stem is —electricity.

It is our intention, hereafter, to treat on these branches, and explain the principle of the electrotype, electric-telegraph, &c. But, as some knowledge of the principles of a science is certainly necessary for the due appreciation of its application—we shall at present confine ourselves to the consideration of electricity: and, as we are not writing for the information of those who are versed in the science, we shall studiously avoid entering into a disquisition on the merits of this or that theory; but restrict ourselves to a plain statement of acknowledged facts. Technical terms, of course, we shall be unable wholly to exclude. So long, indeed, as they are uniformly used in scientific works, it is necessary, perhaps, to familiarize ourselves with them. When using them, however, we shall be careful to accompany them with the requisite explanation.

For many ages past, the science of electricity has been regarded as very important. But, in the present day, this importance is materially increased by the utility which it is found to possess. It is now, in some degree, beginning to be applied, in one form or another, to the arts of life. What eventually may be effected by its aid we dare not trust ourselves to surmise. The day may arrive when even steam, as a moving power, may yield to it in efficiency. Boats have *already* been set in motion by electro-magnetism. With a hundred zinc elements, Stoehrer, of Leipsic, asserts that he believes he can produce a propelling power sufficient to take a train of waggons and passengers by railroad from Leipsic to Dresden, for six shillings: the present cost is five pounds. Thus we shall have a moving power, not only far less dangerous than steam, but infinitely less expensive.

The term "electricity" is derived from elektron—a Greek word for amber:—for the electrical properties of this substance were observed by Thales, a Greek philosopher, so far back as six centuries before the Christian era. Thus the science is of no modern date. Very little respecting it, however, was known before the sixteenth century. Indeed, it was not much before the last century, that any very remarkable facts were developed. About this period the celebrated Benjamin Franklin, among other discoveries connected with the science, identified lightning with electricity.

The electrical properties which Thales observed amber to possess, consisted in the power which it exhibited, after it had been rubbed, of attracting light substances. The ancients consequently supposed, that these electric properties were confined to this substance, and a few others which they afterwards discovered. But modern investigations have proved beyond a doubt, that whatever electricity may be, it exists more or less in every state of the atmosphere, and in all known substances. We say, that whatever electricity may be, since it is not easy to determine the nature of the agent by which electrical phenomena or appearances are produced. It is generally supposed to be a subtle fluid, pervading matter in all its states.

But whatever electricity may be, one thing is certain, it remains balanced and dormant in bodies, until *excited* or roused. Now this excitement does not always result from the same influence. In some bodies the electric fluid is excited by mechanical action—*i.e.*, by rubbing; in others, by the action of heat, magnetism, and by chemical action. Thus, if we take a piece of sealing-wax in our hands, we are not aware of the existence of the fluid;—but if we rub the wax against the sleeve of our coat, or any other woollen substance, the fluid becomes excited, and gives to the wax an attractive power, which is observable when small bits of paper, feathers, &c. are presented to it. The same effects are produced, when a glass tube or phial is rubbed with silk. In these, and similar cases, the disturbing cause is that of mechanical action.

If a magnet be provided with suitable apparatus, both a shock and a spark will be elicited. In this case, the disturbing cause is that of magnetic action.

Again, if heat be applied to resinous or sulphurous substances, electricity is developed during the process of melting. In this case, the disturbing cause is that of the action of heat.

Further, if metals of different kinds be brought into contact (a fluid intervening), electricity is generated. In this case, the disturbing cause is that of chemical action.

We have here, it will be perceived, simply brought forward an example, under the several heads of magnetic, chemical electricity, &c., without going into the detail of explanation. This we shall reserve until we come to the consideration of electro-magnetism — galvanism, &c. At present, we shall confine our attention to that source of electricity, to which we first adverted, viz., to friction or mechanical action. In experiments with the sealing-wax, it will be recollected, that after it had been rubbed, it became endued with an attractive power. It will, however, be found, upon close examination, that another effect, besides attraction, is produced—a totally different effect indeed—we mean repulsion. For, we shall find that the bits of paper, &c. after a momentary adhesion to the wax, are thrown off. These opposite effects may be shewn more clearly, perhaps, by the following experiments.

Take a small apparatus, similar to that in the annexed figure, viz., a stand bent at the top, to which a fine silk thread is attached, having a small pith ball at its end.

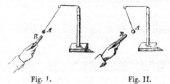

Fig. I.                              Fig. II.

Present a glass tube B, which has been rubbed with silk to this pith ball A: the pith ball is immediately attracted as may be seen by fig. 1; and almost immediately after it is repelled, as observable in fig. 2.

The same result will ensue, if a piece of sealing-wax, rubbed with flannel, be the exciting substance.

Now in the first place we observe that B attracts A; we next perceive that A is repelled from B.

And why is this? Simply because that "bodies when dissimilarly electrified, attract each other, and when similarly electrified, repel each other."

For B, so soon as it is excited, is in a state of active electricity;—while A is in a state of dormant electricity:—hence the attraction being in dissimilar states. No sooner, however, do they come in contact, than B communicates to A a portion of its electricity, and

Figure 3.1 Attraction and repulsion (1846)[117]

---

117    Peck (1846), 125.

Reading in this 'real-time' way was less feasible with the more advanced electrical equipment discussed by Peck, which required detailed instruction, explanations and drawings (see Figure 3.2).

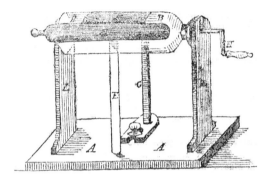

(a) A cylindrical electric machine

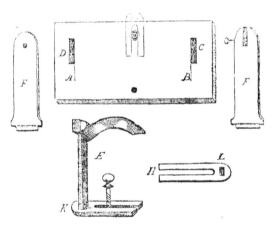

(b) To make the stand for the cushion

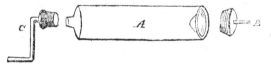

(c) To make the cylinders

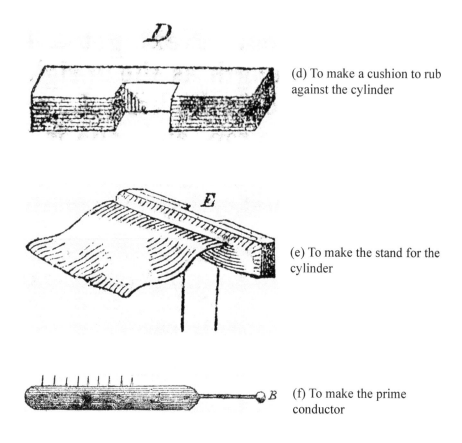

(d) To make a cushion to rub against the cylinder

(e) To make the stand for the cylinder

(f) To make the prime conductor

Figure 3.2 Illustrations of electrical equipment (1847)[118]

Peter Broks argues that, before the nineteenth century, 'readers were rarely invited to participate in science'.[119] However, this seems to be precisely what Peck did. The naivety of the drawings and their close-up perspective indicates that he intended readers to construct the equipment for themselves. At the same time, he solicited their deeper engagement by stating that 'the first question which necessarily proposes itself to us for consideration is, "what is *matter*"' (Peck's

---

[118]    Anthony Peck, 'Papers on Popular Science: II. Electrical Machines', *Reynolds's Miscellany*, 1:9 (2 January 1847), 139–40. The labels given for the further illustrations appear as they were printed in the original.

[119]    Peter Broks, *Understanding Popular Science* (Maidenhead: Open University Press, 2006), 35.

emphasis).[120] He provided the conceptual, practical and visual tools for anyone to make themselves specialists in the subject, regardless of their circumstances, and he supplemented this by offering the information they needed to make electrical equipment, to understand its processes and to conduct their own experiments.

Peck attempted to dismantle the material issues of investigating electricity by giving advice about how to construct a range of electrical equipment. He tried to show not just how electricity works but also how readers might make it work *for them*. Yet readers' own circumstances dictated how the articles were read, in ways that have significant implications for non-fiction science writing. Critics have suggested that difficulties in obtaining appropriate materials made physical sciences, such as chemistry and (presumably) electrical science, more 'aloof' than natural sciences.[121] Certainly, collecting rocks from beaches was easier than making glass cylinders. Equipment such as leather or cutting blades would have been relatively commonplace, with which to follow Peck's instructions for making cylinder cushions: 'now cut a piece of coloured thin leather, one inch longer than the wood, and so wide as to go nearly round it.'[122] To make the actual cylinders, though, Peck advises the reader to 'get a large sample phial, and take care that the sides of it are straight, smooth, and regular'.[123] This would have been considerably more difficult and expensive to accomplish. Very few people at the time had access to scientific laboratories so it would have been a challenge to secure the specified phial. What, then, was the function of Peck's advice? It is unlikely that he thought readers would actually acquire the equipment; they were expected, instead, to follow the described experiments 'virtually' or in their minds. Peck's verbal descriptions and illustrations provided largely imagined scenarios and visualisations of experiments. They would have been read largely for the information they conveyed, rather than as genuine instructions for using equipment, which was what they appeared to be.

While the primary difficulty for specialist scientists lay in grasping the abstractions of electricity beyond the sensory realm, for writers appealing to less specialist audiences, there was the additional difficulty of making the subject understandable. Peck achieves this is by referring to a range of authorities, deploying not just his own knowledge of electrical science or specialist developments but also information from the numerous experimenters who were investigating electricity. It would be anachronistic to describe these reader-experimenters as 'amateurs' for, as David Knight and M.D. Eddy have pointed out, the professionalisation of science

---

[120]   Peck, 'Papers on Popular Science: On Mechanics. I. Matter and its Properties. – Force. – Attraction of Gravitation', *Reynolds's Miscellany*, 2:28 (15 May 1847), 14.

[121]   Amanda Mordavsky Caleb, *(Re)creating Science in Nineteenth-Century Britain* (Newcastle: Cambridge Scholars, 2007), 2.

[122]   Peck, 1:9 (1847), 139.

[123]   Peck, 1:9 (1847).

only began during the nineteenth century.[124] Peck's articles sought to appeal to readers by means of their experiments and contributions to scientific knowledge, rather than formal qualifications or illustrious reputations. In one article alone, he mentions a Mr Armstrong, 'a gentleman of high scientific attainments' who wrote to Faraday; a Mr Ibbertson, who furnished a report to the London Electrical Society; a Mr Pine, 'who has made the electricity of plants his peculiar study'; and a Mr Finlayson 'to whom we are indebted for several facts in connection with this subject'.[125] They provide a cast of background characters in Peck's narrative of experimentation and exemplars whom he encourages readers to emulate. He suggests that 'the chances of profit and reputation are equal and accessible to all' in scientific research might offer his readers 'the way to fame and fortune', and, far from patronising readers, claims that 'the humble and self-manufactured apparatus of the mechanic may achieve discoveries denied to the highly-finished machines of the opulent amateur or the tutored professor'.[126] There is a satirical edge in his use of 'highly-finished', 'opulent' and even 'tutored', in contrast to which his readers' experiments were potentially more authentic, innovative and even superior. His implication that readers were intellectually equal to scientists, with their greater economic or educational privileges, made electrical research not just alluring on an individual basis but also as an implicit reassignment of social, intellectual and economic power.

**Universal Electricity**

Scientists, practitioners and educators shared their interest in electricity through specialist scientific and educational texts, but other forms of popular writing about electricity appealed to different and more generalist interests. As specialist ideas about electricity continued to be refined, they appeared to become increasingly distanced from writings by and for non-specialist readers, particularly in the formalisation of electrical theories through mathematics. As Amanda Caleb suggests, towards the 1870s, the 'segregation of science from the public was reinforced in the professionalization of the sciences'.[127] Generalist periodicals indicate that there was more interest in electricity's practical uses than in its operation and properties, and that comic elements were becoming regular features in responses to the phenomenon.

---

[124]    David Marcus Knight, 'Science and Professionalism in England, 1770–1830', *Proceedings of XIV International Congress of the History of Science, 1974*, vol. 1 (Tokyo, 1975) in *Science and Beliefs: from Natural Philosophy to Natural Science, 1700–1900*, ed. David M. Knight and M.D. Eddy (Aldershot: Ashgate 2005), 53–67.

[125]    Anthony Peck, 'XIII. Thermo-Electricity – Electricity of Steam – Electro-Vegetation', *Reynolds's Miscellany*, 1:27 (8 May 1847), 428.

[126]    Peck, 1:27 (1847), 430.

[127]    Caleb (2007), 2.

In view of the ambivalence already evident towards science and electricity, it would be surprising if there was not some level of satirical response to electrical science. Fiction and non-fiction carried by *Punch* magazine referred repeatedly to electricity and demonstrated a variety of understandings about it; the novelty of electricity meant that it was not only of scientific interest. During the 1840s and 1850s, a multitude of anonymous articles associated electricity with its potentially novel applications, including 'The Electric Parliament', 'Police by Electricity', 'American Electricity', 'Electrical Clocks', 'The Landlord's Electro-biology', 'Our Electric Selves' and 'Cooking by Electricity'.[128] The potential for puns and metaphors was commonly recognised, for example, in the review 'Our Electric Selves' (1854), which discussed a new book about 'electricity and the human body, and the modes of developing it', and the columnist plays with the terms 'shock', 'attraction' and 'repulsion', with reference to both electricity and social behaviour.[129] The laws governing the 'phenomena of mutual attraction and repulsion' are reported to depend on such erroneous factors as hair colour and 'sparks' given off by the eyes. Women are proposed to be the most 'eligible form' by which we might treat 'the body as an electrical machine' because of the power of the hourglass figure to 'electrify' an entire ballroom in an instant.[130] In these jokes, there is no interest in engaging with electricity's scientific composition and only a basic, if any, comprehension of its operation. There is, however, a keen awareness of readers' fascination with the effects of electricity on the body, on society and, indeed, on themselves.

Richard Noakes notes in his study of mid-nineteenth-century science in the magazine that, 'although *Punch* acts as a reliable barometer of the ways in which medical and technological developments were fitting in (or not) into Victorian culture, it does not always give one a sense of the dramatic changes in the "purer" sciences'.[131] Even when Faraday is mentioned in *Punch*, it is in terms of his publicising water pollution or debunking spiritualism. This is not unexpected in

---

[128]   'The Electric Parliament', *Punch*, 8:192 (15 March 1845), 127; 'Police by Electricity', *Punch*, 16:412 (2 June 1847), 225; 'American Electricity', *Punch*, 19:483 (12 October 1850), 160; 'Electrical Clocks', *Punch*, 21:541 (22 November 1851), 228; 'The Landlord's Electro-Biology', *Punch*, 22:570 (12 June 1852), 251; 'Our Electric Selves', *Punch*, 26:661 (11 March 1854), 106; 'Cooking by Electricity', *Punch*, 33:852 (14 November 1857), 198. *Science in the Nineteenth-Century Periodical: An Electronic Index*, v. 3.0, hriOnline [www.sciper.org; accessed 10 July 2010].

[129]   'Our Electric Selves', *Punch*, 26:661 (11 March 1854), 106. There are several books to which the article may refer but the discussion suggests it might be John Obadiah N. Rutter's *Human Electricity: The Means of its Development* (London, 1854).

[130]   See also Alice Jenkins's discussion of how *Punch* satirised the genre of scientific popularisation in, for example, the 1843 article 'Punch's Theory of Light' (Alice Jenkins, *Space and the 'March of Mind': Literature and the Physical Sciences in Britain, 1815–1850* (Oxford: Oxford University Press, 2007), 189).

[131]   Richard Noakes, 'Science in Mid-Victorian Punch', *Endeavour*, 26:3 (September 2002), 95.

terms of *Punch*'s largely non-scientific purpose but, as Noakes also points out, electricity and magnetism 'did matter in *Punch*', even if it was 'only insofar as such forces were used to entertain, improve communication or otherwise improve daily existence'.[132] Indeed, electricity was often welcomed in *Punch* as effusively as elsewhere, for example, when one writer exclaims that 'of all gifts which Science has presented to Art in these latter days, the most striking and magnificent are those in which the agency of electricity has been evoked'.[133] However, *Punch* also provided a distinctly satirical viewpoint, which affected and even determined the sense in which electricity and magnetism were shown to matter.

Despite the relatively minimal development of widespread electrical applications by 1858, one *Punch* columnist had already declared the 'universality of electricity' and envisages the 'good time coming' due to electricity. The future will be, he delights, a time when electricity will 'do everything for us':

> It will cook our dinner, sew on our buttons, write our letters, make our clothes, whip our children, black our boots, shave our stubbly chins, and even help us to a pinch of snuff ... carry us up to bed, undress us, tuck us up, and blow out the candle, when we are too tired, or indifferent to do it ourselves.[134]

Interestingly, no machine is mentioned; instead, it is electricity itself that is personified, as the perfect domestic servant who is capable of any task and never tires. The personification of electrical servitude is echoed elsewhere in the periodical press in the comment, for example, that, 'if electricity, in its wild and natural state, be to man a furious and fitful enemy, it is, when tamed, a patient slave, an obsequious agent'.[135] The *Punch* scenario might appear utopian but, if we look more closely, there is another darker vision of technology than simply 'all the wonder that would be'.[136] The employer of electricity can be seen to have become distanced from his/her responsibilities, effectively infantilised and sapped of energy. The portrayal harks back to Thomas Carlyle's reservations about electricity, in 'Signs of the Times' (1829) that 'not the external and physical alone is now managed by machinery, but the internal and spiritual also'.[137] The human is represented as mere recipient of daily life's habitual conventions and somewhat pointless, as well as reliant on electricity. The depiction stands in marked contrast to the increasing use of electricity as a medical cure for persistent lethargy, rheumatism and hysteria, where patients 'received benefits from' a series of mild

---

[132]    Noakes (2002).
[133]    'The Gifts of Science to Art', *Dublin University Magazine*, 36:211 (July 1850), 3.
[134]    'The Universality of Electricity', *Punch*, 35:902 (30 October 1858), 694.
[135]    'Table Talk', *Once a Week*, 2:34 (22 August 1868), 158.
[136]    Alfred Lord Tennyson, 'Locksley Hall' (1842), l. 120.
[137]    Thomas Carlyle, 'Signs of the Times' (1829), in *The Spirit of the Age: Victorian Essays*, ed. Gertrude Himmelfarb (New Haven: Yale University Press, 2007), 35.

electric shocks.[138] The 'universality of electricity' in *Punch* shows the employer to be invalided by electricity and, as a result, invalidated as his own agent. For contemporary writers and readers, the dangers of electricity were not restricted to the physical impact of shocks; the technological advance of electricity's convenience, its affordability and charm could be as dangerous as it was rewarding. The ambivalence towards electricity and its accompanying technological advances contribute to the negative fiction portrayals I discuss elsewhere here, as well as the dystopian fiction of the later nineteenth century.

New concepts of electricity were frequently applied to practical ends, but the results could be both informative *and* sensational. One writer describes his astonishment at how 'electrizing [*sic*]' a plant seed not only hastened its development but made it germinate with the 'head downwards, and root upwards, in the air'.[139] Another marvels at how electrical currents can 'melt a large iron wire, on passing through it, but which at the same time will pass through the human frame unfelt'.[140] Electricity's unusual qualities prompted writers to speculate on the practical yet equally extraordinary ways in which electricity might extend human powers and fundamentally change living conditions.

The development of electricity contributed substantially to the utopian discourses of the nineteenth century but not consistently so; as Peter Broks points out, 'what was new was their proliferation and popular appeal'.[141] Scientific investigations of electricity revealed the unique and vast potential of its practical applications and the metaphors used to describe them were part of what Broks describes as 'a triumphalist vision of the coming century'.[142] Metaphor was an essential feature in descriptions of an electrical future and yet it also conveyed some of the more problematic aspects of that vision. In 1868, a *World of Science* columnist describes the possibility of warning shipping with electrical buoys along coastlines. He adopts the metaphor of electrical science being as yet an 'infant' and speculates that 'what its manhood will be, no one can imagine, even keeping in view this gigantic first offspring'.[143] The writer converts the intimidating possibilities of electricity to a human scale but, even so, there is something of the monstrous about the gigantic infant he envisages. The ways in which electricity could be simultaneously rewarding and threatening are indicated, too, in the ironic reporting of an invented 'telegastrograph', whereby the flavour of food and drink could be transmitted to the palate from any distance, allowing

---

[138] T.L. Phipson, Dr., 'Electricity at Work', *Macmillan's Magazine*, 6:32 (June 1862), 169.

[139] 'Electricity and Vegetation', *World of Science*, 1:10 (14 December 1867), 140.

[140] 'The Electric Light', *World of Science*, 1:13 (4 January 1868), 181.

[141] Broks (2006), 40.

[142] Broks (2006).

[143] 'The Electric Light', *World of Science*, 1:13 (4 January 1868), 182.

enjoyment at no cost but also providing no nutrients.[144] Faced with such unknown yet clearly extraordinary possibilities, metaphors like this indicate that, while there was excitement about electricity, there was also something disturbing about the uncertainty of its future development.

The 'cultural embeddedness' of science is demonstrated by the range of topics nineteenth-century writers relate to electricity.[145] Electricity was viewed by readers in terms of scientific development and technological applications, but it also symbolised a number of issues relating to the future progress of the human race. Scientific investigations of electricity provided the understandings necessary for the development of new electrical technologies but they did so in the context of several other factors. As Thomas Hughes points out, 'technological affairs contain a rich texture of technical matters, scientific laws, economic principles, political forces, and social concerns'.[146] These links were simultaneously edifying and entertaining, rather than separate as Morus suggests.[147] Although Hughes focuses on late nineteenth-century electrical distribution systems rather than responses to electricity or its representation, his observation that 'electrical power systems embody the physical, intellectual, and symbolic resources of the society that constructs them' is also relevant to writings about electricity before the 1880s.[148]

New concepts about physical matter caused a radical upheaval of Victorian perceptions about the natural world; however, their significance is less recognised than that of other nineteenth-century conceptual revolutions in theories of, say, evolution or geology. As we saw earlier from the concerns expressed by Faraday, Maxwell and Whewell, complex and abstract ideas about physical matter did not lend themselves easily to the narrative forms by which information could be conveyed. Those who wrote about electricity were also not necessarily knowledgeable practitioners; writers responded to, transformed and fictionalised concepts of electrical power, within a range of forms. The article, 'What is Electricity?' (1861) is of particular interest because it engages with ideas about electricity at sufficient length and depth to reward further scrutiny. The article was published in the two-penny, illustrated weekly *Once a Week* and reveals both public perceptions and inherent conflicts about electricity and its nineteenth-century development.[149]

---

[144] 'More about Electricity', *Chambers's Journal of Popular Literature, Science and Arts* 790 (15 February 1879), 108.

[145] Geoffrey Cantor, Gowan Dawson, Richard Noakes, Sally Shuttleworth and Jonathan R. Topham, 'Introduction', in Henson et al. (2004), xvii.

[146] Thomas P. Hughes, *Networks of Power: Electrification in Western Society, 1880–1930* (Baltimore: Johns Hopkins University Press, 1983), 1.

[147] Morus (2007), 2.

[148] Hughes (1983), 2.

[149] 'What is Electricity?' *Once a Week*, 4:84 (2 February 1861), 163–5. *Once a Week* (London: Bradbury and Evans, 1859–1880) started with a price of two pence; unfortunately, prices do not appear on later editions.

The article's author begins by posing the question that continued to trouble nineteenth-century public perceptions of electricity: 'what definitely am I to think of when I say that word?'[150] His initial response is to review eighteenth-century understandings of electricity, by mentioning the Leyden jar, Benjamin Franklin, Humphry Davy and De la Rive, and then briefly describing the new processes of electro-chemistry and electrolysis.[151] However, despite these illustrious and empirical authorities, he is not satisfied. He suggests that it is insufficient to recognise simply that electricity is not two fluids and that it is a single force; as he states 'but, "what kind of force?" is still the question'.[152] Unusually, in view of his original question, he does not mention the work of more recent scientists involved in electrical science or physics. Instead, he acknowledges that nineteenth-century investigations of electricity are barely more than 'fanciful speculations' and that the new theories are not 'established truth'; for him, they merely constitute a hypothesis that 'seems best to harmonise and bind together a great body of anomalous facts' and which will 'stand or fall' as knowledge increases.[153] The importance of establishing fine distinctions is clear from his account and his review, like the new science he discusses, makes it distinct from more whimsical conjectures.

After the writer's candid presentation of the technical and historical background of electrical science, he turns to what he sees as the repercussions of electricity's new role in the world. He starts by drawing attention to how investigations of electricity have fundamentally altered man's understanding of existence itself. Old and new understandings are imaged in terms of substance, with solidity and certainty being replaced by the new fast-moving fluidity of questions and revelations: now, he says, 'we see the firmest unions dissolved, the elements in definite proportions carried this way and that, and forced into new combinations'.[154] The changes are not entirely positive though; as he suggests, revelations about physical matter introduce a degree of confusion about life whereby nothing seems certain anymore. There is a palpable sense of the disorientation such radical discoveries prompted when he compares the extent of change in perceptions to those brought about by astronomy, which revealed that the earth was not comfortably 'slumbering on its broad foundations, but hung baseless mid infinity, "it taketh no rest"'.[155] The expression paraphrases a biblical quotation that describes man's anxious disquiet, in the face of his own vain and futile existence.[156]

---

[150] 'What is Electricity?' (1861), 163.
[151] 'What is Electricity?' (1861), 163–4.
[152] 'What is Electricity?' (1861), 164.
[153] 'What is Electricity?' (1861).
[154] 'What is Electricity?' (1861).
[155] 'What is Electricity?' (1861), 165.
[156] Ecclesiastes 2.22–23: 'For what hath man of all his labour, and of the vexation of his heart, wherein he hath laboured under the sun?/For all his days are sorrows, and his travail grief; yea, his heart *taketh not rest* in the night. This is also vanity' (my emphasis).

It conveys the off-kilter perspectives of a time that had also yet to recover from the implications of Charles Darwin's *On the Origin of Species*, published just two years before. Warming to his theme, the article's author suggests that perhaps people have been 'equally deceived at the opposite end of the scale' and that the conception of matter as solid, with relatively stationary particles, 'may be overthrown'; indeed, if 'ceaseless motion proved the condition of existence for atoms as for worlds. What then?'[157] There is a hint of alarm and desperation about the climactic two-word question. To contemporary writers, electrical research was clearly disconcerting; as another contemporary commentator remarks, electrical science was 'altering the relations between the empire of man and the worlds of time and space'.[158] Despite the rewards offered by the development of electricity, it also demanded deep emotional and psychological adjustments, in the face of nineteenth-century scientific concepts that we now take for granted.

Even successes in understanding electricity could prompt reservations. As the *Once a Week* columnist points out, 'substances that had baffled all other means of chemical analysis and were regarded as elementary, electricity has resolved'. He portrays electricity as more than a tool for discovery; it is also an agent of discovery. In his depiction, electricity assumes a similar degree of agency to the perfect servant in the *Punch* portrayal. He suggests that man's senses are unreliable and that he would be unwise to consider sensory evidence as 'impregnable ground'.[159] In contrast, scientific investigation has the rigour of a lawyer interrogating the integrity of the senses: 'science deals with them as an able lawyer deals with a pack of stupid or roguish witnesses: cross-questions them, sets one against the other, sifts and balances conflicting evidence, marshals it, puts sense into it, – and in the end triumphantly draws truth out of it'.[160] The metaphors of the battleground and the courtroom reflect a marked antagonism between empirical science and the new scientific processes of investigation. However, the writer also defends the 'stupid and roguish' senses, claiming that 'we cannot afford to despise our senses, since through them alone comes our report of the world without … It is but shallow philosophy to sneer at the senses, for without them man's reason would be a king without a kingdom'.[161] Even as the author celebrates the 'ingenuity' of science, he presents the senses as vital to man's rationalism, without which there would be no point to his nature as a reasoning being.

The article's conclusion describes how science enhances man's capacities, in ways that are both attractive and disturbing. The telescope, for example, makes man's eyes 'the eyes of a giant' and the microscope allows him to see into 'the mysteries of the smallest flower, like King Oberon himself'.[162] Man's

---

157    'What is Electricity?' (1861).
158    'The Origin of Electricity', *Saturday Review*, 17:438 (19 March 1864), 348.
159    'The Origin of Electricity' (1864), 164.
160    'The Origin of Electricity' (1864), 165.
161    'The Origin of Electricity' (1864).
162    'The Origin of Electricity' (1864).

technological capabilities are figured as affecting the nature of humanity itself, and the new electrical equipment as ideal for capturing the elusive nature of 'Nature's stealthiest agent' – that is, electricity.[163] He reveals the confusion and disquiet prompted by investigations of the physical sciences and particularly electricity, aspects which are often neglected in histories of nineteenth-century science. As suggested previously, in my introduction, electricity and increasing knowledge about it appeared to exemplify a growing disjunction between man and the natural world. While the phenomenon's power and potential were acknowledged, its intimidating, mystical and even furtive qualities persisted, and were also identified by fiction authors as core characteristics of the phenomenon.

The objections to electricity expressed by these periodical writers are not the same as those of Wordsworth or those described by Harriet Ritvo as 'preservationists', whose tone she describes as 'nostalgic, emotional, and evocative of aesthetic value'.[164] No pastoral ideal was envisaged or defended, but electricity was portrayed as invasive and simultaneously appealing and damaging; indeed, part of its menace was its appeal and vice versa. The disturbing revelations about the nature of physical matter, inherent in electrical research, took place against the questioning of other long-held certainties, such as biblical authority, grand historical narratives and social hierarchies. Little wonder, then, that the anonymous columnist of *Once a Week* appeared so disoriented or the *Punch* writer so sceptical. The interrogation of what was 'real' shook the basis of non-fiction as a form of representation for, if reality was suddenly so uncertain, perhaps any attempt to represent it realistically was also futile.

The difficulty of representing electricity was an important factor in its attractiveness as a subject for literary representation, whether the author's purpose was explanatory or metaphorical. Electricity occupied a similarly intriguing and 'contested space' to that which Iwan Rhys Morus allocates machinery.[165] Like the increasingly sophisticated machinery being developed during the 1830 to 1880 period, electricity could be controlled, at least potentially, and offered the ability to control what was beyond one's self. What made electricity uniquely alluring, though, was its invisibility – and that gave it a secret quality. The secret nature of electricity appears to underlie its continuing allure and offers a key to understanding the differences that appear in writings about it. Different writers sought to share the secrets of electricity yet revelled equally in the enigmas and paradoxes of electricity; their responses are also reactions to man's understandings of and relationship to the worlds within and around him.

---

[163]   'The Origin of Electricity' (1864).

[164]   Harriet Ritvo, 'The View from the Hills: Environment and Technology in Victorian Periodicals', in Henson et al. (2004), 168.

[165]   Morus (1998), 156.

# Chapter 4
# Electrical Shorts

The period from the 1830s to the 1880s was a time of growing excitement, uncertainty and awareness about electricity, despite what might sometimes have appeared to be the 'stumbling progress' of electrical investigation.[1] During the same period, the cheap periodical emerged as the era's most popular form of reading and became a regular venue for short-fiction responses to electricity. The conceptual scope of ideas about electricity expanded within the heterogeneous content of the periodical form and in response to the new demands of changing readerships, giving fiction authors the freedom to explore the subject as a literary and imaginative force, as well as a science, so that it gained entirely new associations. Fiction in British periodicals remains comparatively unexplored; however, the popularity of nineteenth-century periodicals alone makes the fiction they carried important, and particularly as a literary response to electricity.

Recent understandings of the relationship between nineteenth-century periodical literature and contemporary cultures have altered significantly and rapidly, in ways that directly influence the study of writings about electricity and their relationship to contemporary literary and scientific cultures. In the mid-1980s, it was thought that popular science periodicals 'mirrored above all editors' concerns, aspirations, and prejudices'.[2] However, the model of print cultures 'mirroring' society can be questioned on the basis that any 'mirrored' reflection only ever presents a one-way image. In fact, print cultures constituted 'a central component of that culture' and printed material 'can only be read and understood as part of that culture and society'.[3] Rather than writings in periodicals occupying a marginal position in relation to the concerns of literature, the topics they address are increasingly understood as expressions of contemporary cultures. For that reason, the employment of periodicals as a primary form is no longer peripheral to the study of either science or writing but, instead, a unique and crucial element of it. In N. Katherine Hayles's 'feedback loop' between literature and science, just as in Beer's 'two-way' traffic, cultural developments are in continual interchange with

[1]  John Desmond Bernal, *Science and Industry in the Nineteenth Century* (London: Routledge, 2006), 130.

[2]  Susan Sheets-Pyenson, 'Popular Science Periodicals in Paris and London: The Emergence of a Low Scientific Culture, 1820–1875', *Annals of Science*, 42:6 (November 1985), 552.

[3]  Lyn Pykett, 'Reading the Periodical Press: Text and Context', in *Investigating Victorian Journalism*, ed. Laurel Brake, Aled Jones and Lionel Madden (Basingstoke: Macmillan, 1990), 7.

the cultural innovation.[4] The actual content of scientific knowledge was altered dramatically by what authors and readers understood and, just as importantly, what they failed to understand or chose to understand from entirely different angles.

The growing appreciation of wider publication forums as literary resources has prompted intensive examination of nineteenth-century readerships, the press and popular scientific knowledge, in relation to culture, media and commerce.[5] What emerges from this body of work is that, in the first half of the nineteenth century, a remarkable fusion occurred between scientific development, literary contexts, institutional reform and private commerce. Four decades of factory acts and reformist legislation ushered in a rise in literacy rates at much the same time as the 1836 reduction in paper taxes and stamp duties which, in turn, enabled the production of cheaper publications.[6] More information was available to more people than ever before, and the years in which this phenomenon reached its zenith were precisely when the development of professionalised scientific knowledge began to flourish. It would seem then that the transformation of print culture, readerships and society was intimately connected to Britain's commercial growth, as well as the democratisation and professionalisation of the sciences. The dynamism of the process and its ramifications is particularly evident in the periodical press, which was effectively the fastest moving and most responsive publication arena of the time.

The production and consumption of scientific knowledge through different forms of literature is especially pertinent to the study of electrical writings, because nineteenth-century electrical science was practised throughout such a broad range of backgrounds and levels of expertise. Some participants were scientists but others were entertainers, educators or simply curious members of the public. Meanwhile, the writings they produced were distributed to and by an equally diverse range of non-specialist reading publics and publication methods. The periodical writings examined in this volume are intended to highlight the fact that many impressions of science were recorded only in the fragmentary ephemera of scientists' letters and lecturer's asides, the anecdotal fictions that feature

---

4    N. Katherine Hayles, *Chaos Bound: Orderly Disorder in Contemporary Literature and Science* (Ithaca: Cornell University Press, 1990), xiv; Beer (1983), 5.

5    Peter Allan Dale, *In Pursuit of a Scientific Culture: Science, Art and Society in the Victorian Age* (Madison: University of Wisconsin Press, 1989); Louise Henson, Geoffrey Cantor, Gowan Dawson, Richard Noakes, Sally Shuttleworth and Jonathan R. Topham, eds, *Culture and Science in the Nineteenth-Century Media* (Aldershot: Ashgate, 2004); Laurel Brake and Julie F. Codell, eds, *Encounters in the Victorian Press: Editors, Authors, Readers* (Basingstoke: Palgrave Macmillan, 2005); Aileen Fyfe and Bernard Lightman, *Science in the Marketplace: Nineteenth-Century Sites and Experiences* (Chicago: University of Chicago Press, 2007); James Mussell, *Science, Time and Space in the Late Nineteenth-Century Periodical Press: Movable Types* (Aldershot: Ashgate, 2007).

6    The 1801 Health and Morals of Apprentices Act; 1819 Factory Act; 1831 (Hobhouse) Factory Act; and 1847 Factory Act; the stamp duty reductions of 1836 were followed by their final abolition in 1855.

in instructional writings, isolated short stories and the metaphors employed by novelists. The variations between these forms were not simply a matter of literary genre or style; they arose from fundamentally different aims, which shaped their representation in literature. Changes in the consumption of scientific writing were accompanied by equally radical changes in the practice and purposes of science. By the 1840s, contrary to Matthew Arnold's definition of science as 'knowledge systematically pursued and prized in and for itself', science was no longer such an organised, gentlemanly or insular pursuit. As Alan Rauch has pointed out in *Useful Knowledge* (2001), the allure of science was both macro-economic and micro-economic; scientific progress was a matter of national pride, reflecting growing British prosperity, as well as the country's desire to compete as an international power and devise potentially exciting technological advances. However, it also offered the individual the possibility of '*mental improvement*' (Rauch's emphasis) and, with that, the hope of elevated social status.[7]

Despite the appearance of more available, frequent and diverse accounts of scientific practice and knowledge, as disciplines became more specialised, new distance occurred between expert knowledge and 'common' knowledge, and between scientific language and vernaculars. In her study of nineteenth-century electrical technologies, *When Old Technologies were New* (1988), Carolyn Marvin points out that 'in scientific and technical literature, expert authority rejected the immediate sensory judgement, or direct experience of nature, as naïve empiricism'.[8] Advances in science meant that Baconian empiricist approaches were no longer sufficient for specialist scientists; interacting directly with complex scientific concepts demanded that more abstract, theoretical approaches be devised. In contrast, Marvin suggests, popular science 'appealed to the senses [and] referred to a way of knowing quite unlike the controlled observational posture of empirical investigation'.[9] What this indicates is that the producers of scientific knowledge were motivated by very different interests than those of non-specialist consumers. As a model, it asserts two opposed 'ways of knowing', where a superior 'controlled' observational empiricism is preferable to the 'naïve' and sensory, ostensibly random nature of popular science. These two 'ways of knowing' were not always so starkly opposed, though. In Faraday's experimentation as well as James Clerk Maxwell's poems, for example, controlled observation could also be naïve and sensory.

The range of available approaches to science in the nineteenth century indicates that learning was not always the purpose of science writings, and that the more diverse readerships were, the less predictable intentions were. Some writers sought a degree of understanding about electricity and energy sciences but others, regardless of their educational or social status, sought entertainment too.

[7]  Alan Rauch, *Useful Knowledge: The Victorians, Morality, and the March of Intellect* (Durham, NC: Duke University Press, 2001), 24.

[8]  Carolyn Marvin, *When Old Technologies were New: Thinking about Electric Communication in the Late Nineteenth Century* (Oxford: Oxford University Press, 1988), 111.

[9]  Marvin (1988), 109.

As a result, those who wrote about science had to do more than just interpret or simplify specialist scientific knowledge: they had to offer multiple narratives to engage newly demanding and disparate audiences.

## Science and Fiction: Debatable Ground

The liveliness of nineteenth-century exchanges between culture, literature and forms of print was a defining influence in the development of popular awareness about scientific ideas. As Tim Killick suggests, the vibrancy of the interactions is also especially apparent in the magazines of the period.[10] The nineteenth-century serialisation of novels in the periodical press continues to enjoy considerable scholarly attention.[11] However, apart from a few notable exceptions, there has been relatively little study of short fiction in periodicals, let alone the way its authors engaged with scientific topics.[12] This neglect ignores a vast body of material that was a primary reading source for Victorian readers, but references to short periodical fiction still tend to be limited to study guides on the short story or, more recently, anthologies.[13] Recent scholarship on periodical fiction examines literatures either from before the 1830s, such as Gillespie et al.'s *Romantic Prose Fiction* (2008), or from after the 1880s, such as Winnie Chan's *The Economy of the Short Story in British Periodicals of the 1890s* (2007).[14] While there is also extensive scholarship on short fiction, it too tends to begin from 1880 or to focus on writings by non-British or female authors.[15] Although these studies are

---

[10]    Tim Killick, *British Short Fiction in the Early Nineteenth Century: The Rise of the Tale* (Aldershot: Ashgate, 2008), 19. Killick's study is primarily of short fiction between 1800 and the 1830s.

[11]    For example, Jennifer Hayward, *Consuming Pleasures: Active Audiences and Serial Fictions from Dickens to Soap Opera* (Lexington: University Press of Kentucky, 1997) or Graham Law, *Serializing Fiction in the Victorian Press* (Basingstoke: Palgrave Macmillan, 2000).

[12]    Geoffrey Cantor and Sally Shuttleworth, eds, *Science Serialized: Representation of the Sciences in Nineteenth-Century Periodicals* (Cambridge, MA: MIT Press, 2004).

[13]    Leslee Thorne-Murphy, 'Students Researching Victorian Short Fiction', *Academic Exchange Quarterly*, 10:1 (Spring 2006), 232–6. Recent studies of the short story include: Andrew Maunder, *The Facts on File Companion to the British Short Story* (New York: Infobase Facts on File, 2007); Charles Edward May and Frank Northen Magill, *Critical Survey of Short Fiction: Essays, Research Tools, Indexes* (1981; Salem Press, 2001).

[14]    Gerald Ernest, Paul Gillespie, Manfred Engel and Bernard Dieterle, *Romantic Prose Fiction* (Amsterdam: John Benjamins, 2008); Winnie Chan, *The Economy of the Short Story in British Periodicals of the 1890s* (London: Routledge, 2007).

[15]    Relevant examples include: [post-1880] Cheryl Alexander Malcolm and David Malcolm, *A Companion to the British and Irish Short Story* (Oxford: Wiley-Blackwell, 2008); [non-British] Edward W.R. Pitcher, *An Anthology of the Short Story in 18th and 19th Century America* (Lewiston and Lampeter: E. Mellen Press, 2000); Peter Cogman,

not considered in depth here, I refer to them to indicate the considerable gap in scholarship that exists and which this chapter seeks to address. The examples of short fiction I consider were published between 1838 and 1884, a period in which, as discussed previously, ideas about electricity were at their least stable. The period undertaken is considerably wider than that of comparable studies on nineteenth-century literature and periodicals, such as Dallas Liddle's *The Dynamics of Genre: Journalism and the Practice of Literature in Mid-Victorian Britain* (2009), which concentrates on the 1850s and 1860s and also focuses exclusively on fiction.[16] Both fiction and non-fiction are studied in the present work, in order to compare the conventions that arose between the different forms and to illustrate the vital exchanges that took place between them, as neglected areas of scholarship.

The content of short fiction in the nineteenth century is directly related to its struggles for literary status and negative assessments of its literary worth. The demand for short fiction altered significantly between 1800 and 1900 but its cultural location is still an ongoing subject of debate. As Harriet Devine Jump suggests, what was once known as the 'tale' escaped the limitations of its original market within 'low' or popular genres, and became part of the greater 'aesthetic and cultural prestige' of the short story.[17] In contrast, Cheryl and David Malcolm propose that contemporary readers continued to favour the novel and particularly the serialised novel, with the result that short fiction remained 'undervalued and scarcely reflected on by its practitioners and readers'.[18] The value contemporary writers and readers accorded to short fiction is impossible to gauge properly, though, without considering three important contextual factors. Firstly, short fiction was published predominantly in the periodical and was associated, therefore, with what was essentially a popular and relatively ephemeral forum. Secondly, short fiction in cheap periodicals may have held considerable greater value to readers with limited resources in terms of finance, leisure time and literacy, for whom it frequently constituted the only form of literature to which they had access. Thirdly, assessments of the extent to which short fiction was 'reflected on' are inherently flawed by the absence of readers' opinions, which were largely under-represented or unrecorded. Considerable further scholarship would need to be conducted on short fiction in the periodical and its readers before its role in contemporary

---

*Narration in Nineteenth-Century French Short Fiction: Prosper Mérimée to Marcel Schwob* (Durham: University of Durham, 2002); Alfred Bendixen and James Nagel, *A Companion to the American Short Story* (Oxford: Wiley-Blackwell, 2010); [women] Harriet Devine Jump, *Nineteenth-Century Short Stories by Women: A Routledge Anthology* (London: Routledge, 1998); Christine Palumbo-DeSimone, *Sharing Secrets: Nineteenth-Century Women's Relations in the Short Story* (Madison: Fairleigh Dickinson University Press; London: Associated University Presses, 2000); Margaret Beetham and Kay Boardman, *Victorian Women's Magazines: An Anthology* (Manchester: Manchester University Press, 2001).

[16]   Dallas Liddle, *The Dynamics of Genre: Journalism and the Practice of Literature in Mid-Victorian Britain* (Charlottesville: University of Virginia Press, 2009).

[17]   Devine Jump (1998), 1.

[18]   Malcolm and Malcolm (2008), 8.

readers' lives can be assessed with any accuracy. One consequence of cursory judgements of nineteenth-century short fiction is the modern perception that it lacks literary merit, and its dismissal as 'unoriginal' at best and 'patronising' at worst.[19] The resulting critical neglect of the form represents a significant gap in literary scholarship, which is perhaps exacerbated by the reluctance to 'let go of the book' in favour of periodicals noted by Amy King.[20] However, as Killick points out, to treat any era as 'one of relative infertility' effectively also 'marginalises a large part of the history of the genre'.[21] If we move beyond judgements based on current literary criteria, short fiction in periodicals can be understood as a certain type of literary practice, which deliberately seeks to engage with other literatures and concerns in a form beyond the canonical text.

To call short fiction before mid-century a 'genre' would be somewhat artificial and anachronistic, as the boundaries of many literary genres had yet to be established. However, in order to discuss responses to electricity in periodical fiction further, we need to identify a stable terminology. Although Edgar Allan Poe claimed in 1842 that 'the Tale has peculiar advantages which the novel does not admit', in the first half of the nineteenth century, as Killick points out, 'the distinction between tale, sketch, essay, and so forth was not as clear-cut as that which modern criticism would impose'.[22] Short fiction was an altogether messier affair, in accordance with the characteristic 'messiness' scholars have identified in many nineteenth-century art forms.[23] Rather than occupying a neatly bordered category of its own, short fiction was more usually associated with a variety of other 'ephemeral and elastic' modes of writing, such as comic extracts, single-volume tales, narrative essays and sketches.[24] In periodicals, it was as nebulous and 'fragmentary' as the forum itself which, as James Mussell suggests, gestured to a range of spaces.[25] In my discussion, I defer to Killick's argument that 'short fiction' conveys an appropriate indistinctness and avoids the more theoretically loaded associations of 'short story'.[26] However, where terms such as 'story' or 'tale' improve fluency or clarity, I employ these too.

---

[19]   Gowan Dawson, Richard Noakes and Jonathan R. Topham, 'Introduction', in *Science in the Nineteenth-Century Periodical: Reading the Magazine of Nature*, ed. Geoffrey Cantor, Gowan Dawson, Graeme Gooday, Richard Noakes, Sally Shuttleworth and Jonathan R. Topham (Cambridge: Cambridge University Press, 2004), 18.

[20]   Amy King, 'Searching out Science and Literature: Hybrid Narratives, New Methodological Directions, and Mary Russell Mitford's *Our Village*', *Literature Compass*, 4 (2007), 1485.

[21]   Killick (2008), 5, 7.

[22]   Edgar Allan Poe, 'Twice-Told Tales: A Review', *Graham's Magazine* (April 1842), reprinted in *Edgar Allan Poe: Essays and Reviews*, ed. G.R. Thompson (New York: The Library of America, 1984), 568; Killick (2008), 22.

[23]   See David Trotter, *Cooking with Mud: The Idea of Mess in Nineteenth-Century Fiction* (Oxford: Oxford University Press, 2000).

[24]   Killick (2008), 19.

[25]   Mussell (2007), 61.

[26]   Killick (2008), 10.

In the fiction considered here, responses to electricity are closely coupled with concerns about electrical experimentation, and concepts of conductivity, fluidity and electrical circuitry are particularly prominent. Before the middle of the century, understandings of battery operations were often sketchy and, even after mid-century, confusion about electrical theories appears to dominate. Although scientists and other practitioners attempted to root contemporary investigations of electricity in the well-informed results of experiments, as my discussion of scientific conceptualising indicates, the principles of electricity were equally ruled by hypotheses. In a period when even specialists appeared unable to represent or articulate consistent facts about electricity, fiction authors relied on their own interpretations and responded to electricity by means of their various understandings. In nineteenth-century fictions about electricity, facts are rarely correct; however, unlike non-fiction writings, they were designed to entertain, rather than disseminate accurate information. Writings about electricity occupied what, in 1845, the lecturer Abraham Hume dubbed 'the debatable ground between science and fiction'.[27] The ideas about electricity disseminated by means of fiction were unequivocally interpretations and speculations but, even as such, they were not necessarily so distant from the hypothetical speculations of contemporary science.

When short fiction in periodicals is discussed, it is rarely examined in depth. Here, five examples are investigated, which span the period between the 1830s and 1880s, as follows:

1. Richard Johns, 'Mr. Hippsley, The Electrical Gentleman', *Bentley's Miscellany* (1838).[28]
2. Anon., 'Reminiscences of a Medical Student', *New Monthly Magazine and Humorist* (1842).[29]
3. Anon., 'The Tree of Knowledge', *Dublin University Magazine* (1853).[30]

---

[27]   Abraham Hume, 'Remarks on the Vestiges of the Natural History of Creation', *Liverpool Journal*, 1 February 1845, 5; quoted in James A. Secord, *Victorian Sensation: The Extraordinary Publication, Reception, and Secret Authorship of Vestiges of the Natural History of Creation* (Chicago: University of Chicago Press, 2000), 230. In this instance, Hume was voicing objections to the *Vestiges of Natural Creation* (1844). I am grateful to James Secord for confirming that the figure to whom he refers is Abraham Hume (1814–1884), who was a lecturer in English Literature at the Liverpool Mechanics Institute and the Collegiate Institution in Liverpool.

[28]   Richard Johns, 'Mr. Hippsley, The Electrical Gentleman', *Bentley's Miscellany*, 4 (July 1838), 374–82.

[29]   'Reminiscences of a Medical Student', *New Monthly Magazine and Humorist*, 64:253 (January 1842), 123–41. The title may be an oblique reference to the third chapter of Thomas Carlyle's *Sartor Resartus*, which is also entitled 'Reminiscences'.

[30]   'The Tree of Knowledge', *Dublin University Magazine* 41:246 (June 1853), 663–75.

4. Richard Dowling, 'An Anachronism from the Tomb', *Tinsley's Magazine* (1879).[31]
5. Anon., 'Doctor Beroni's Secret', *Temple Bar* (1884).[32]

The stories have been chosen for several reasons, the first being that each seeks to portray electricity's processes and effects, as well as characters who experimented with them. This makes them unusual, as electricity tends to be referred to in periodical fiction either briefly or metaphorically, a response that also underlies many of the other writings. In this chapter, I seek to demonstrate the variety of fiction responses to electricity, which ranges from comic ephemera to elaborate tales of the supernatural. For clarity, the stories are grouped according to the key concepts by which they relate to one another. Each story emphasises a different aspect relating to electricity, the sum of which provides an informed and useful overview of contemporary impressions. They are discussed in terms of how they engage with ideas of electricity, and they have been chosen from each decade during the period of study to reflect how responses to electricity changed in short fiction over time. The early nineteenth century has been described as 'an extraordinary period in British cultural history', because public discourse about science and literature and the relationship between the two was conducted within a 'uniquely broad range of social and intellectual contexts'.[33] The stories discussed in this chapter illustrate the changing nature of the relationship and contexts, and how they influenced writings about electricity. I focus on investigating and comparing the stories, rather than the nature of the publication forums; however, the examples were also chosen because they were published in demonstrably popular periodicals, making the ideas they expressed available to relatively wide and non-specialist readerships and potentially influential. For transparency, the prices and circulation figures for the publications are provided below (see Table 4.1).[34]

Whether considered as a group or individually, the stories may not ever be regarded as esteemed contributions to the literary canon, but they do represent the way in which electrical science and experimentation simultaneously emerged out of and reflected contemporary interests and anxieties. Their shortness and ephemerality as literary forms is also important, as part of the less sustained reading methods which are increasingly recognised to have been a significant,

---

[31]    Richard Dowling, 'An Anachronism from the Tomb', *Tinsley's Magazine*, 24 (February 1879), 147–66.
[32]    'Doctor Beroni's Secret' (I), *Temple Bar*, 72 (October 1884), 187–208; 'Dr. Beroni's Secret' (II), *Temple Bar*, 72 (November 1884), 339–60.
[33]    Alice Jenkins, *Space and the 'March of Mind': Literature and the Physical Sciences in Britain, 1815–1850* (Oxford: Oxford University Press, 2007), 18.
[34]    *Waterloo Directory*, online edition [www.victorianperiodicals.com/series2; accessed 28 May 2011].

even dominant, feature of Victorian reading practices.[35] The unifying factor between the stories is their engagement with the topic of electricity, which offers some characteristic 'unity of impression'.[36] More directly, the purpose of studying them is to explore what Gowan Dawson describes as the addition of 'new and unexpected connotations to the original scientific ideas'.[37] In that sense, direct correlations between the fictions and electrical sciences are explored here when they arise, but the aim is also to show what further thinking was prompted by ideas about electricity. Rather than approaching the writings as productions of either literature or science, the aim is to show the ways in which writings about electricity were more frequently what Amy King describes as 'hybrid forms – neither strictly literary nor strictly science' but, rather, varieties of both at the same time.[38]

Table 4.1 Prices and circulation figures for selected periodicals

| # | Short story title | Periodical name | Price | Circulation per issue (monthly) |
|---|---|---|---|---|
| 1 | Mr. Hippsley, The Electrical Gentleman | *Bentley's Miscellany* (1838) | 2s 6d | 6,000 (1837) to 8,500 (1839–1840) |
| 2 | Reminiscences of a Medical Student | *New Monthly Magazine and Humorist* (1842) | 3s 6d | 5,000 (1830) to 3,000 (1860) |
| 3 | The Tree of Knowledge | *Dublin University Magazine* (1853) | 2s 6d | 4,000 (1844) to 3,000 (1860) |
| 4 | An Anachronism from the Tomb | *Tinsley's Magazine* (1879) | 1s | 10,000 (1870) |
| 5 | Doctor Beroni's Secret | *Temple Bar* (1884) | 1s | 15,000 (c.1884) |

Before considering the individual stories, it is important to establish the significance of the periodical publication context of the short fiction discussed and define the terminologies I have used in approaching them, which I seek to achieve in the following section. In doing so, I also interrogate some of the assumptions

---

[35] Nicholas Dames, *The Physiology of the Novel: Reading, Neural Science, and the Form of Victorian Fiction* (Oxford: Oxford University Press, 2007).

[36] Edgar Allan Poe suggests that the quality of short fiction depends upon creating a 'unity of impression' that can achieved only if a tale is read in one sitting ('The Philosophy of Composition' in *The Oxford Book of American Essays*, ed. Brander Matthews (New York: Oxford University Press, 1914), 101).

[37] Gowan Dawson, 'Science and its Popularization', in *The Cambridge Companion to English Literature, 1830–1914*, ed. Joanne Shattock (Cambridge: Cambridge University Press, 2010), 172.

[38] King (2007), 1491.

about short fiction that have shaped recent scholarship, as well as the original reception of the stories.

### Electrified Men

Metaphors for electricity were especially persistent in short fiction. Indeed, it is possible that the brevity of the form makes metaphor especially useful because, as one of the most influential scholars of the short story, Charles May, suggests, brevity means that each 'detail is transformed into metaphoric significance'.[39] With the condensed nature of shorter fiction, as with poetry, literary methods that convey multiple meanings and nuances, simultaneously and succinctly, are particularly valuable. The transformative potential of metaphor is evident in the first fictions I consider: Richard Johns's 'Mr. Hippsley, The Electrical Gentleman' (1838) and Richard Dowling's 'An Anachronism from the Tomb' (1879), both of which respond to notions of electricity, particularly conductivity, the fluid analogy and electrical circuitry.[40]

The earlier story, 'Mr. Hippsley' (1838), is a comedy and it was published in the monthly *Bentley's Miscellany*, just when the magazine was nearing the peak of its circulation (see Table 4.1). The eponymous Charles Hippsley believes that he is so charged with electricity that he is a danger to others. As a result, he has broken off his engagement with his fiancée, Catherine, and is increasingly distraught. We understand that he is electrified from the protagonist's presentation of his condition and the images he adopts from contemporary science. He describes himself as a 'walking electrical machine' and a 'walking galvanic battery'.[41] By the 1830s, the concept of electrical conductivity was already well established, having been announced in 1732 by the experimental naturalist, Stephen Gray.[42] In Gray's eighteenth-century experiments on conductivity, he examined various materials including 'silk cords' for their quality as conductors.[43] 'Mr. Hippsley' appears to be informed by a version of this basic understanding of conductivity, supplemented by an awareness of new battery technology. When Hippsley arrives at a meeting with Catherine, for example, it is noted that Catherine's 'dress, and even her gloves, were

---

[39]    Charles E. May, 'Why Short Stories are Essential and Why They Are Seldom Read', in *The Art of Brevity: Excursions in Short Fiction Theory and Analysis*, ed. Per Winther, Jakob Lothe and Hans H. Skei (Columbia: University of South Carolina Press, 2004), 18.

[40]    Johns (1838); Dowling (1879).

[41]    Johns (1838), 374.

[42]    Michael Ben-Chaim, *Experimental Philosophy and the Birth of Empirical Science: Boyle, Locke and Newton* (Aldershot: Ashgate, 2004), 25. Gray's discovery was followed in the early 1730s by the work of his colleague Granvil Wheler, and of Charles Dufay from the Académie Royal des Sciences, and the Royal Society's curator John Desaguliers.

[43]    Ben-Chaim (2004), 27.

of silk, [and are] consequently non-conductors of electricity'.[44] Writing in the late 1830s, Johns's understanding of the non-conductivity of silk was probably gained also from more recent sources than Gray, such as Edward Turner's *Elements of Chemistry: Including the Recent Discoveries and Doctrines of the Science* (1835).[45] Turner refers repeatedly to the non-conductivity of silk, in reports of electrical experiments, and the volume was 'for some years the text book used by almost all teachers'.[46] Indeed, the book is still considered to have been 'one of the best of all nineteenth-century textbooks of chemistry'.[47] Although Johns incorporates actual and specific aspects of conductivity in his story, he also adapts the concept and exploits his readers' misunderstandings for their comic potential.

The contemporary image of the electrified body engaged with real-world referents that were simultaneously conceptual, physical and social. Johns combines the body-as-battery analogy with the electrical fluid analogy in his narrator's depiction of 'the subtle electric fluid coursing through the muscles', an image of man that perhaps related to the eighteenth-century 'Electric Boy' parlour games discussed previously.[48] The metaphor worked both ways, with the battery described as an invention that 'requires but little *food*, and with that little will perform an honest day's work, whether it be to suspend a weight, bring platina wire into a red heat, or keep up a powerful rotary motion' (author's emphasis).[49] The analogy was not restricted to fiction and sometimes acted as the basis for popular explanations of how the body worked, for example, in an 1839 columnist's description of how 'the human body, in a natural state of health, not exhausted by fatigue nor depressed by cold, has a perceptible + electricity' and 'the − electricity of the exhalation'.[50] As a conceptual aid, the idea of man as an electrical being persisted well into the century. In 1850, the chemist and surgeon, Arthur Smee, described an 'electro-biological doctrine' that 'man acts by electricity, which is set in motion through the muscular structures', an assertion he illustrated with images of the skin, ear, eye and muscle connecting to the brain

---

[44]   Johns (1838), 378.

[45]   Edward Turner, *Elements of Chemistry: Including the Recent Discoveries and Doctrines of the Science* (Edinburgh: William Tait and Charles Tait, 1827).

[46]   'Edward Turner, M.D., F.R.S., Obituary, *Gentleman's Magazine* (April 1837), 434.

[47]   W.H. Brock, 'Turner, Edward (1796–1837)', *Oxford Dictionary of National Biography* (Oxford University Press, Sept. 2004; online edn, January 2009) [http://ezproxy. ouls.ox.ac.uk:2117/view/article/27848, accessed 10 July 2010]. Turner was Professor of Chemistry at University College, London, and authored several papers in scientific periodicals and in the Transactions of the Royal Societies of Edinburgh and London. Brock reports that Turner's book went through eight editions, being revised and enlarged after his death by his brother, W.G. Turner.

[48]   Johns (1838), 379.

[49]   'Odds and Ends', *John Bull*, 854 (24 April 1837), 94 (author's emphasis).

[50]   'Electricity of Animal Life', *The Penny Magazine of the Society for the Diffusion of Useful Knowledge*, 8:439 (2 February 1839), 48.

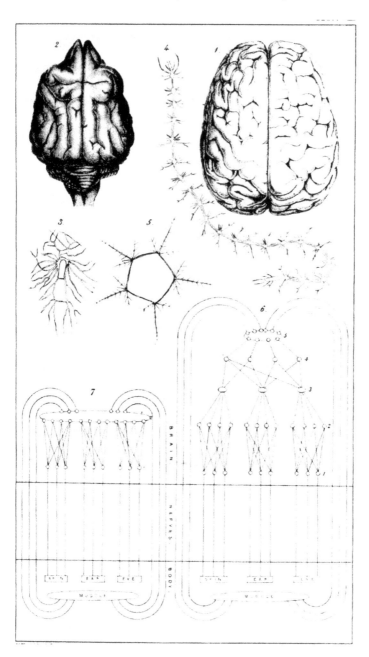

Figure 4.1 Arthur Smee, *Instinct and Reason* (1850)[51]

---

[51]    Smee (1850), 210. Smee's accompanying explanation suggests that the human body has lower and higher functions, such as sensation, action and will. Brain and nerve

along wire-like nerves (see Figure 4.1).[52] Mechanical and electrical metaphors for the body's operation were so common and so persuasive that, even a decade after Smee, scientists such as George Henry Lewes were still protesting about how 'misleading' it was, to think of 'the brain as a galvanic battery of which the nerves are conducting wires'.[53] With the uncertainty about both bodily sensations and the nature of electricity, the two were easily superimposed and provided a ready supply of easily available metaphors.

A further aspect of Hippsley's existence as an electrified man is offered in his description of himself as a 'human *upas*' tree, a reference that amalgamates the electrical fluid analogy with the wider contemporary scientific interests of botany and exploration.[54] The 'upas', a deciduous tree, was reported to exist in Africa and Asia in the late eighteenth century, as well as the mid-1830s.[55] The tree yielded latex, which was used by local hunters as an arrow poison, and it was obtained by a puncture or incision being made to release the yellowish-coloured sap. Mr. Hippsley's comparison of himself to one of these trees is intended to convey the poisonous nature of his electrified condition, but, in imaging a plant whose essence is liquid, it can also be read as an underlying extension of the fluid analogy for electricity.

The comic potential of being electrified was fully recognised by late 1830s and 1840s periodical writers. However, literary engagements with scientific concepts were often determined by publication forums and readerships as much as authors' understandings of scientific developments. In 1836, just two years before the publication of Johns's tale, John Frederic Daniell devised the wet-cell battery.[56] In the story, it is the concept of an electrical circuit which provides both the complication and the resolution of the story, as well as Mr Hippsley's condition, and it is adapted to suit the purposes of the fiction. When Hippsley falls from his horse, he is told that his arm has been broken and that 'the fracture in your frame has destroyed that unity of parts, that wonderful sympathetic combination which

---

fibres, labelled 3, 4 and 5, illustrate the similarity between the two.

[52]   Arthur Smee, *Instinct and Reason: Deduced from Electro-Biology* (London: Reeve and Benham, 1850), 212, 216.

[53]   George Henry Lewes, *The Physiology of Common Life*, vol. 2 (New York: Appleton, 1867), unnumbered footnote, 20.

[54]   Johns (1838), 379.

[55]   The 'upas' tree *(Antiaris toxicaria)* was first reported to exist in Java in the 1780s by N.P. Foersch, a Dutch East India Company surgeon, but his claim was later discredited. The tree was documented for the first time in the anonymous article 'Upas, The Poison Tree of Java', *Encyclopaedia Britannica*, vol. 17 (1823), 58, and, later, in the *London and Edinburgh Philosophical Magazine and Journal of Science*, vol. 6, ed. David Brewster, Richard Taylor and Richard Phillips (London: Taylor & Francis, 1835), 218.

[56]   Sir William Grove developed the 'gaseous' galvanic cell battery later in 1839; see Gilbert M. Masters, *Renewable and Efficient Electric Power Systems* (London: Wiley-IEEE, 2004), 208.

had rendered you an electrical phenomenon'.[57] The oblique reference to electrical circuitry is as technical as the tale gets; *Bentley's Miscellany*, in which the story appears, is known to have steered clear of 'specifically scientific and religious matters'.[58] The author is not entirely naïve about scientific ideas though; as the narrator comments, 'the doctor knew he was talking nonsense [to Hippsley], but looked wonderously grave to conceal this fact' and throughout the story everyone but Hippsley knows he is deluded.[59] A similar degree of irony is evident in the *Punch* article 'The Electrical Minister', which proposes experiments to prove Prime Minister Robert Peel's 'powers of attraction and repulsion', because he is 'quite as powerful as an electric eel'.[60] What Richard Johns's story indicates is that he and contemporary readers were sufficiently acquainted with electrical concepts to laugh off the notion of someone being an 'electrical gentleman'. It was not necessary for readers to understand the technical or accurate details of scientific concepts to establish a coherent literary response; the intention was only that they understood the references sufficiently to understand their literary function.

The later short fiction 'An Anachronism from the Tomb' (1879) by Richard Dowling engages with further aspects of electrical circuitry and fluid analogies.[61] Despite the analogies perpetuating misunderstandings about the nature of electricity, the images they portrayed continued to be familiar and in common use, not least, as we have already seen, by Faraday and Maxwell. In the story, a scientist called Byron Favell attempts to create 'frozen electric fluid', although, as he points out, it is 'a contradiction in terms, but a convenient manner of expressing your meaning'.[62] The author and narrator make it clear that the analogy is not a literal explanation and is used somewhat ironically. Favell succeeds in freezing 'two cubic feet' of electricity, using mercury mixed with 'two hundredweight of frozen oxygen in cubes and one hundredweight of frozen chlorine in pellets'.[63] However, unlike the narrative undercutting of Mr Hippsley's electrical delusion, Favell's feat is made to appear as convincing as possible.

The account of Favell's electrical experiment might appear to be based on mere fantasy but certain aspects engage closely with contemporary science. In 1877, just two years before, gases with low critical temperatures such as oxygen were successfully liquefied by the French physicist Louis Paul Cailletet (1832–1913)

---

[57]    Johns (1838), 382.

[58]    Logan Browning, 'The Irregular Publication of "Regular Habits": Dr. Charles Julius Roberts and "Bentley's Miscellany"', *Victorian Periodicals Review*, 23:2, Wellesley Index Special Issue (Summer 1990), 60.

[59]    Johns (1838), 382.

[60]    'The Electrical Minister', *Punch*, 10 (28 March 1846), 145.

[61]    There is no definite biographical information about Dowling; however, he may be the Irish novelist Richard Dowling (1846–1898), whose romance novel was published in the same year by the Tinsley Brothers, proprietors of *Tinsley's Magazine*.

[62]    Dowling (1879), 147.

[63]    Dowling (1879), 154.

and Swiss chemist Raoul Pierre Pictet (1846–1929), independently.[64] Cailletet and Pictet made their discovery known by telegram to John Tyndall and it was clearly a matter of some excitement, as its announcement in the London *Times* indicates.[65] Explorations of cryogenic methods began in the 1820s, with Faraday's success in liquefying gases with high critical temperatures and the properties of the elements and gases mentioned in the story, such as mercury, oxygen and chlorine, were still under investigation.[66] In the story, there is an awareness of the methodological obstacles presented by low temperatures, if not complete understanding. We are told that 'the thick walls of the Leyden jar shrank to the thinness of tissue-paper' before Favell succeeds in manufacturing a 'globule' of electricity, which he handles with long tongs and carries 'suspended by a fine silk thread, and in a current of ether spray'.[67] The term used to describe Pictet's method was 'the cascade principle'.[68] The fictitious Favell, meanwhile, achieves a comparable 'brisk shower of electricity in a liquid form'.[69] How widely circulated reports of Pictet's work were is hard to establish, but a degree of resemblance emerges between the non-fiction and fiction writings. *Tinsley's Magazine*, in which the story was published, was based in London and, assuming that the author was too, it is not unlikely that he had at least heard about *The Times*' report of the development and adopted the terminology used to describe it.[70] Even if the author was not aware of Pictet's metaphor, the language of both chemist and author appears to be similarly inspired by the fluid analogy.

The cryogenic processes about which Dowling writes were at the cutting-edge of nineteenth-century scientific endeavour and they were understood properly by only very few specialists. Nevertheless, fictions such as 'An Anachronism from the Tomb' do not refer only to specific, factual developments; they also convey the cultural framework of contemporary scientific innovations. They envision the scientific processes, the people involved and the applications to which discoveries might be put. By considering this wider framework within which fiction writings about science operated, we remain open to the 'greater plurality of the sites for the making and reproduction of scientific knowledge', as Cooter and Pumfrey suggest.[71] In short periodical fiction, inaccuracy does not necessarily detract from

---

[64]    Anthony Kent, *Experimental Low Temperature Physics* (Basingstoke: Macmillan 1993), 9. See also, J.L. Heilbron and James Bartholomew, *The Oxford Companion to the History of Modern Science* (Oxford: Oxford University Press, 2003), 474.

[65]    'Liquefaction of Oxygen', *The Times*, 25 December 1877, 4.

[66]    See Michael Faraday, *On the Liquefaction and Solidification of Bodies Generally Existing as Gases* (London: R. and J. E. Taylor, 1845).

[67]    Dowling (1879), 155, 156.

[68]    Steven W. Van Sciver, *Helium Cryogenics* (New York: Plenum, 1986), 4.

[69]    Dowling (1879), 155.

[70]    'Liquefaction of Oxygen' (1877).

[71]    Roger Cooter and Stephen Pumfrey, 'Separate Spheres and Public Places: Reflections on the History of Science Popularization and Science in Popular Culture', *History of Science*, 32 (September 1994), 254.

authenticity; instead, it conveys the further contexts of scientific exploration, the excitement of its revelations, and the possibilities of its applications. Rather than seeking direct representations of science, these aims allow us, perhaps, to be somewhat more forgiving of the scientific elements in 'An Anachronism from the Tomb', not to mention the tale's increasingly absurd progression.

Electricity and experimentation gather further ghoulish associations, in which the liquid electricity soon becomes an agent for the eerily unnatural. As Martin Willis suggests, since the time of Isaac Newton 'a confusing range of opinions' existed about whether electricity was a solid or a fluid and the fluid theory itself was 'indicative of the mystical importance of electrical force'.[72] Favell uses the frozen electricity in conjunction with a small galvanic battery, to resurrect a corpse from a neighbouring cemetery, inserting the globule into its mouth and then applying the battery to the topmost vertebra that connects the skull and spine, 'causing the current to pass from a point as close as possible from the atlas to the heel of the right foot'.[73] The use of electricity to revive the corpse relates the fiction to actual contemporary experiments with the medical uses of electricity, even as it alludes to earlier associations with electrical reanimation in the Vitalism debate, the experiments on cadavers by Galvani and Aldini discussed previously, and Mary Shelley's *Frankenstein*. The fictitious corpse turns out to be that of Oliver Goldsmith, who was, in fact, buried in central London, as the story suggests.[74] His body is found intact, having been in suspended animation as a result of catalepsy, and it is 'suddenly loosed from that frozen condition by means of a galvanic spark'.[75] Although the story was published long after Goldsmith's death, it appears to be an extended pun on the author's *An History of the Earth, and Animated Nature* (1774), which was published in the last year of his life. The volume was so popular that it went through over 20 editions, which might itself be deemed a perpetual renewal of life, about which the short story is also perhaps an implicit joke.[76] Once revived, Goldsmith fails to persuade nineteenth-century publishers to publish his work and, this being more than he can stand, he commits suicide by deliberately making himself the conductor in an electrical circuit, seizing 'the fatal wire' of a battery and bringing his foot swiftly down on another.[77]

---

[72]    Martin Willis, *Mesmerists, Monsters, and Machines: Science Fiction and the Cultures of Science in the Nineteenth Century* (Kent, OH: Kent State University Press, 2006), 68.

[73]    Dowling (1879), 160.

[74]    Oliver Goldsmith (1728?–1774) was buried at Temple Church, while the cemetery in the story is located at an apparently fictitious 'St. Bridget's Street'.

[75]    Dowling (1879), 160.

[76]    In Goldsmith's *History*, he discusses atmospheric and static electricity and the 'electric fluid' but he never refers to electricity as a means of reviving the dead.

[77]    Dowling (1879), 166. The story is unlikely to be connected to death by electrocution, which was not introduced until 1888 in New York. For information about the introduction of electrocution as a form of capital punishment, see Paul Finkelman, *Encyclopedia of American Civil Liberties*, vol. 1 (New York: Routledge, 2006), 244.

Electricity is given a central function in what is, otherwise, little more than entertainment and at points the author does refer to real contemporary scientific developments and possible future electrical applications. However, the narrator makes clear his awareness of the ludicrous basis of the fiction and the way in which the story unfolds is laden with irony. The author's reluctance to engage fully or seriously with the scientific elements of electricity accords with the growing perception that fictional engagements with science somehow diminished narrative authority or credibility. To read electricity as solely scientific would be to deny its other equally important facets; however, the phenomenon can be characterised partly as scientific, and writings about it may have been influenced by the lack of literary authority and canonicity that later presided over science fiction. The relationship between writings about electricity and science fiction are discussed more fully elsewhere in this text; however, it can be noted at this point that certain literary engagements with electricity resemble the early forms of science fiction that, as Tom Shippey proposes, constituted a 'threat' to 'conservative groups' – in both literature and science.[78] By inference, they would then operate like satire, as ways of broaching established social, scientific and literary boundaries.

Scientific themes have a specific role and character in short stories; however, the self-containment of the short story form may also have acted as a form of censorship on scientific complexity. The short story secures readers' attention only briefly and even for a single instance, unlike serialised fiction to which readers repeatedly return to follow the discussion of a continuing theme or topic. As I mentioned earlier in this chapter, the form's limitations have been suggested by Charles May as a reason for the predominance of the metaphor, as a conceptual short-cut. Scholars acknowledge that serialised fiction 'incorporated theories of mind or exploited metaphors derived from botanical taxonomy or energy physics', and there is no reason to suppose that the same would not apply to individual short fictions.[79] The stories considered so far respond to ideas about conductivity, fluidity and circuitry, but they do so indirectly and without attempting to offer explanations. The short story's uniquely fleeting nature, in terms of how it was read, may also contribute to its characteristic engagement with multiple, overlapping and seemingly disparate interests. Effectively, short stories published in nineteenth-century periodicals were widely cast nets, aiming to capture the broad interests of readers as quickly as possible.

---

[78]   Tom Shippey, 'Literary Gatekeepers and the Fabril Tradition', in *Science Fiction, Canonization, Marginalization, and the Academy*, ed. Gary Westfahl (Santa Barbara, CA: Greenwood Publishing Group, 2002), 8.

[79]   Geoffrey Cantor and Sally Shuttleworth, 'Introduction', in Cantor and Shuttleworth (2004), 2.

## Experimenters and Experiments

A particularly recurrent connection that emerges in relation to electricity is with unbalanced psychological states, and not just in relation to the medical uses of electricity that pre-dated the 1830s.[80] It is apparent from the fictional characters affected by electricity discussed here, that electricity involved scientific obsession and neurosis. The association is immediately apparent if we review the psychological states of the characters endowed with electricity, firstly, in the stories already discussed. Mr Hippsley, the 'wretched hypochondriac', is also described as 'a maniac'; during his meeting with Catherine, he 'paced the room in a state of excitement bordering on frenzy'; and, at the end, everyone is delighted 'at the restoration of the invalid to a natural state of mind', with the implication that his previous state was 'unnatural'.[81] In 'An Anachronism from the Tomb', Oliver Goldsmith becomes increasingly 'exasperated and depressed' and has 'a fit of the spleen', until ultimately he commits suicide in distress at his literary failure.[82] In each case, electricity is portrayed as causing a level of physical excitability that leads to psychological abnormalities.

The association with electrical excitement as an abnormal condition appears to draw on the early adoption of the term 'excited' by electrical science, which was first employed in relation to electricity in the seventeenth century by Robert Boyle in a literary comparison of 'seraphic love' with the electrostatic attraction between two needles.[83] By the late eighteenth century and early nineteenth century, the metaphor of excitement features commonly in references to electrified bodies in contemporary scientific dictionaries.[84] The way in which short fictions share the metaphor of excitement can be viewed as a form of engagement with contemporary electrical sciences, but only on a relatively superficial level. What makes the engagement by fiction more profound is the personification of electrical excitement. By humanising the metaphor, fiction authors expand its possibilities and use it to explore other related contemporary interests, such as the causes of psychological imbalance, its social implications and the curative potential of new technologies.

Substantial scholarship already exists on the use of electricity as a medical treatment in the nineteenth century and its concerns lie beyond the immediate

---

[80]   See, for example, Michael La Beaume, *Remarks on the History and Philosophy but Particularly on the Medical Efficacy of Electricity in the Cure of Nervous and Chronic Disorders* (London: F. Warr, 1820).

[81]   Johns (1838), 376, 378, 382.

[82]   Dowling (1879), 164.

[83]   Robert Boyle, Hon., *Some Motives and Incentives to the Love of God, in a Letter to a Friend*, 4th edn (London: printed for Henry Herringman, 1665).

[84]   See, for example, Charles Hutton, *A Philosophical and Mathematical Dictionary* (London: printed for the Author, 1815), and James Mitchell, ed., *Dictionary of the Mathematical and Physical Sciences* (London: printed for Sir Richard Phillips, 1823).

scope of the current study.[85] Of greater relevance here is the suggestion that contact with electricity could cause psychological problems, through excessive exposure, experimentation or scientific interest. As Tim Killick suggests, during the nineteenth century, short fiction ceased to be associated solely with 'sentimental romance, simplistic allegory, and explicit moral didacticism' and, instead, began to share the novel's concerns with psychological and social realism, as well as its aspirations to artistic and historical credibility.[86] In the short fiction examined here, responses to electricity are often inseparable from responses to other sciences, especially those undergoing similarly rapid development during the same period. The interconnectedness of nineteenth-century electricity with the emerging sciences of psychology and botany is particularly interesting in the two anonymous short fictions 'Reminiscences of a Medical Student' (1842) and 'The Tree of Knowledge' (1853).[87] Both stories focus on electrical experimentation but the other sciences mentioned also feature prominently enough to help determine the portrayal of both electricity and those who were involved with it.

Although 'Reminiscences of a Medical Student' was published in an ostensibly humorous journal, it is more darkly entertaining than comic. It tells the story of Elias Johns, a fellow medical student at Guy's Hospital of the now elderly narrator.[88] The narrator mentions that they studied together during the time of 'the Hunters, Franklin, Watt, Lavoisier, [and] Jenner', figures who were most active in science between the 1760s and 1780s.[89] In reading 'Reminiscences' as a response to electricity, we need to keep in mind that the narration and authorship straddle two vastly different periods, as far as electricity is concerned. Its 1842 publication reaches back to eighteenth-century scientific explorations and places them alongside the most significant early nineteenth-century discoveries about electricity. In doing so, the tale is effectively a palimpsest of impressions on the subject.

In the figure of Elias Johns, a possible association emerges between religious non-conformity and electrical experimentation. Elias is described as 'a most

---

[85]   George M. Eckert, Felix Gutmann and Hendrik Keyzer, *Electropharmacology* (Boca Raton: CRC Press, 1990); Iwan Rhys Morus, *Bodies/machines* (Oxford: Berg Publishers, 2002); Linda Simon, *Dark Light: Electricity and Anxiety from the Telegraph to the X-Ray* (Boston: Houghton Mifflin Harcourt, 2005); Iwan Rhys Morus, 'Bodily Disciplines and Disciplined Bodies: Instruments, Skills and Victorian Electrotherapeutics', *Social History of Medicine*, 19:2 (2006), 241–59.

[86]   Killick (2008), 6.

[87]   Anon. (1842); Anon. (1853).

[88]   Guy's Hospital, London (f. 1721) was at the leading edge of anatomical sciences in the early nineteenth century. For more information about the early history of Guy's, see S.J. Peitzman, 'Bright's Disease and Bright's Generation – Toward Exact Medicine at Guy's Hospital', *Bulletin of the History of Medicine*, 55:3 (1981), 307–21.

[89]   The scientists mentioned are John Hunter (1728–1793); William Hunter (1718–1783); Benjamin Franklin (1706–1790); James Watt (1736–1819); Antoine Lavoisier (1743–1794); and Edward Anthony Jenner (1749–1823).

singular being' – an eccentric but brilliant medical student who is obsessed with studying electricity, in order to revitalise corpses.[90] He is usually found to be alone, although 'his manner was most winning' and he was 'a desirable friend'.[91] The narrator emphasises that, despite the name 'Elias Johns' sounding Jewish, the character is not.[92] The comment not only affirms that he is not Jewish but also leads the reader to speculate on what Elias's origins might be. We are told at the outset that Elias is 'science mad' and that 'his particular hallucination was electricity, with its collaterals, galvanism, and the sciences of heat and light'.[93] The portrayal of electrical experimentation as a form of monomania is explicit; however, more subtle associations are also apparent. As Susan Tucker shows in her study of the semantic development of the term 'enthusiasm', it was increasingly associated with dissenters.[94] It is worth noting, too, that in the late nineteenth century the surname 'Johns' belonged almost exclusively to residents of Wales and Cornwall, regions that were also predominantly Non-Conformist.[95] Paul Wood and Geoffrey Cantor establish in their study of Michael Faraday as a 'Sandemanian and scientist' that there was also a particular tradition of scientific and electrical experimentation among dissenters.[96] In 'Reminiscences', Elias Johns's disastrous electrical experimentation is part and parcel of his unhinged dissenting, in what seems a derogatory comment on both features, as well as a form of warning to readers.

While there is virtually no technical detail in the story, Elias's belief in the unity of electrical phenomena echoes Faraday's theories. The narrator characterises Elias as a misguided materialist and, significantly, his fixation with electricity is shown to border on blasphemy within his portrayal as a dissenter. Although the fictional account draws on earlier ideas of electrical reanimation and vitalism, the technology Elias envisages also resembles defibrillation technology, which would not be demonstrated until 1899 at the University of Geneva.[97] Elias 'worshipped his electrical deity' and believed that

> The electric fluid was the God of Nature, – that the human soul, and all other intelligences were but modifications, but portions of this principle, and at

---

[90]    Anon. (1842), 123. I refer to the character by his first name to avoid confusion with Richard Johns, author of 'Mr. Hippsley', the story discussed previously.

[91]    Anon. (1842), 124.

[92]    Anon. (1842), 123.

[93]    Anon. (1842), 125.

[94]    Susie I. Tucker, *Enthusiasm: A Study in Semantic Change* (Cambridge: Cambridge University Press, 1972).

[95]    Great Britain Family Names Profiling, University College London [http://gbnames. publicprofiler.org/Surnames.aspx; accessed 2 September 2011].

[96]    Geoffrey Cantor, *Michael Faraday: Sandemanian and Scientist: A Study of Science and Religion in the Nineteenth Century* (Basingstoke: Macmillan, 1991); Paul Wood, *Science and Dissent in England, 1688–1945* (Aldershot: Ashgate, 2004).

[97]    P.R. Fleming, *A Short History of Cardiology* (Amsterdam: Rodopi, 1997), 194.

death returned to it again. That it pervaded the universe, was the cause of all phenomena – the source of every change in matter – the creator of worlds, and the chain of systems.[98]

The 1842 authorship of the story pre-dates the establishment of a British school of practical chemistry, which started in London in 1845 as the Royal College of Chemistry.[99] So, although Elias is from a poor background, he has to pay for all his own equipment to be made, in order to continue experimenting. It is his father who brings him funds and, when he does, there is a manic aspect to Elias's joy; indeed, he 'appeared completely possessed' and 'his eye burned with a wild enthusiasm'.[100] Meanwhile, his faith in electricity has hints of egomania; as the narrator tells us, '"Give me", was a favourite sentence of his, "give me boundless space, matter in atoms, Electrical Attraction and Repulsion, and I will soon create you a universe!"'[101] The exclamation modifies the Cartesian declaration, 'give me matter and motion, and I will construct the universe' – the ultimate materialist reduction of existence. Although the original statement might have suited Elias's goals just as well, the crazed new version demonstrates his mania and it is electricity that constitutes the core tension.

Eventually Elias's experiment succeeds and very grandly so, culminating in the public resurrection of a hanged criminal at a large anatomical theatre. Up to this point, the plot of 'Reminiscences' may have been inspired by the real case of the Reverend Dodd (1777), whom the surgeon and anatomist, John Hunter, tried unsuccessfully to resurrect after he had been hanged, using respiration instead of electricity – a case that also seems to have inspired Dickens's short story 'The Black Veil' (1836). In 'Reminiscences' though, as soon as the corpse rises, Elias realises to his horror that it is his own father, who has been committing robberies to pay for his son's experiments and equipment. Before we have time to consider the scenario's comedy or moral message, Elias falls into 'a fit of catalepsy or some anomalous nervous affection', a condition of unconsciousness and bodily rigidity.[102] He dies after experiencing a curious feeling, footnoted in the text as the *aura epileptic*, a physical warning sensation, characteristic 'to those afflicted with epilepsy and other nervous disorders when a fit is about to come on'.[103] There has been no mention previously of Elias suffering from epilepsy but, throughout the tale, his state of mind has been questioned. He is described as being unusually

---

[98]   Anon. (1842), 125.
[99]   The Royal College of Chemistry was originally based on Oxford Street in central London. It operated between 1845 and 1872, after merging with the Royal School of Mines in 1853. It was the first constituent college of Imperial College, London and eventually became the Chemistry Department there.
[100]   Anon. (1842), 132.
[101]   Anon. (1842), 126.
[102]   Anon. (1842), 137.
[103]   Anon. (1842), 139.

pale and thin, with eyes that have 'an absent, wild, dreamy, mystic sort of an expression'; 'he took food as he did sleep, by snatches, quick and hurried, reading as he ate' and 'even when he walked about, he was continually calculating or scheming'.[104] Taken as a whole, the depiction suggests Elias's propensity towards nervous disorder, which combines with his obsessive study of electricity to cause his final collapse and death.

On a deeper level, 'Reminiscences' responds to electricity as an underlying feature of connections between the body, brain and behaviour, and it was published at a time when understandings of epilepsy and mental illness were undergoing critical development. In ways that pre-date late nineteenth-century medical uses of electricity, the portrayals of Elias and his death indicate the embedded nature of electricity in contemporary medicine and psychiatry.[105] By the nineteenth century, the *aura epileptic* had long been recognised as a genuine medical phenomenon.[106] From 1857, the terms 'positive' and 'negative', so commonly used in electrical science, began to be used for referring to levels of activity, excitability and behaviour in epileptic and schizophrenic conditions.[107] The foundations were established for references to electricity in relation to later medical understandings and terms for epilepsy.[108] By the 1870s, behaviour was firmly linked to mental states, on the basis that inhibitory centres were more 'powerful' in some individuals than others, and that 'one of the earliest signs in many cases of insanity is a

---

[104]   Anon. (1842), 123, 127.

[105]   See, for example, G. Beard and A. Rockwell, *On the Medical and Surgical Use of Electricity* (New York: William Wood and Company, 1891).

[106]   The *aura epileptic* ('aura', *L.* and *Gr.* breath, breeze; 'epilepsy', *L.*, to take hold of) was originally documented by Galen (c.130–210AD), whose writings on epilepsy were translated by several authors in the early twentieth century. (See, for example, Mervyn J. Eadie and Peter F. Bladin, eds, *A Disease Once Sacred: A History of the Medical Understanding of Epilepsy* (Eastleigh: John Libbey 2001), 24.) See also the seminal text on epilepsy by Owsei Temkin, *The Falling Sickness: A History of Epilepsy from the Greeks to the Beginnings of Modern Neurology* ([1945]; rev. 1971; Baltimore: Johns Hopkins University Press, 1994).

[107]   German E. Berrios, 'Positive and Negative Symptoms and Jackson: A Conceptual History', *Archives of General Psychiatry* 42:1 (January 1985), 95. According to Berrios (1985), the terms were first used in the context of psychiatric conditions by John Russell Reynolds (1828–1896) in the paper 'On the Pathology of Convulsions' (1857), delivered to the North London Medical Society. The terms were employed further in 1875 by John Hughlings Jackson (1835–1911) to describe the example of a deluded patient who thinks his nurse is his wife (positive element), which relies on his *not* knowing that she is his nurse (negative element), an association between 'not knowing' and 'wrong knowing' (J.H. Jackson, 'The Factors of Insanities', *Med Press Circular*, 108 (1894), 617).

[108]   ESES (epilepsy with electrical status epilepticus) is the modern term for epilepsy with continuous spikes and waves during slow sleep. See Joseph Roger, *Epileptic Syndromes in Infancy, Childhood and Adolescence, International Workshop on Childhood Epileptology* (London: John Libby, 1985), 265–84.

diminution in the inhibitory power of these centres'.[109] In 'Reminiscences', Elias's obsessive behaviour originates in his innate lack of inhibition but, ultimately, it is his misguided and excessive exposure to electricity that debilitates and finally destroys him.

The catalepsy Elias suffers was often associated with hysteria, as it was described in detail in the medical journals and press of the day.[110] It was also thought that 'persons of a nervous temperament' were susceptible to catalepsy, especially when faced with conditions of 'violent anger, protracted grief, hatred, and sudden terror' or 'long continued and intense mental application'.[111] Elias's mania represents an excess of 'positive' psychological symptoms, which find expression in the multi-layered electrical metaphor. His brilliance is represented by his edginess and his obsession with electricity and, in the father's resurrection, the revitalising electrical shock is transferred psychologically and physically to the son. Elias aimed to achieve an electrically resurrected life but, instead, he becomes the conductor in the circuit between life and death.

The obsession with electricity is also the downfall of the narrator of 'The Tree of Knowledge' (1853).[112] Again, the narrator is an old man but this time he is called Melchior, who is not only 'brilliant', 'eloquent' and 'extraordinary' but, we are told, his research 'seemed to have exhausted every branch of human knowledge'.[113] Like the narrator of 'Reminiscences', Melchior tells a story of youthful obsession with electricity but this time it is his own story. As a young man, Melchior goes to study in Germany and one day, when he is experimenting with the effects of positive and negative electrical currents, he notices that the marks left by electricity bear 'a strange and very striking resemblance to the foliage of a tree'.[114] From this point, he notices the pattern everywhere, including in the frost on windows, and begins to think that electricity is the source of everything.[115] What he describes as 'the demon desire of knowledge' leads him to think that if he can harness the pervasive power of electricity, he might discover 'the original cause and germ of vegetable life'.[116] Like Elias Johns, he shuts himself away to pursue his experiments, 'literally intoxicated' by the creative potential electricity might

---

[109]   W.M. Rutherford, 'Lectures on Experimental Physiology: Lecture IV: Innervation', *The Lancet* 1 (1871), 566; Rutherford also quoted in Berrios (1985), 96.

[110]   See, for example, 'Dr. Duncan on "Catalepsy"', *The Lancet*, 2 (London: Elsevier, 1830), 277–84.

[111]   John Eberle, *Treatise on the Practice of Medicine*, vol. 2 (Philadelphia: Grigg and Elliot, 1835), 73.

[112]   Anon. (1853), 663–75. Despite its name, the magazine's circulation was not restricted to Dublin University; it was published between January 1833 and December 1877 and, in 1853, circulation was between 3,000 and 4,000 issues per month (WDENP).

[113]   Anon. (1853), 675.

[114]   Anon. (1853), 666.

[115]   We might note also the correlation between these ideas and Arthur Smee's electrical images in 1850 (see Figure 4.1).

[116]   Anon. (1853), 667.

give him; however, unlike 'Reminiscences', the first-person narration encourages us to share his experiences.[117] Melchior thinks of creating a new plant species using electricity, an idea that quickly acquires a separate and relentless identity. The obsession is conveyed as follows:

> It filled my thoughts by day, my dreams by night; it never left me time for food or relaxation; it haunted me like a familiar; in the street, in the lecture room, in the fields, in my own chamber, wherever I moved or rested, it was for ever with me, and whispering to me.[118]

The personification of Melchior's idea mimics the schizophrenic delusion of a separate entity and completely supplants his earlier affection for Margaret, the daughter of his professor. He becomes so 'besotted' with his work that he suffers from the paranoid delusion that she and her father are colluding to distract him from his great discovery. He builds himself a glass chamber in which to conduct his experiments, from materials that were 'best adapted to the influence of the electrical laws'.[119] At one point, he is found raving, full of 'anguish and madness', oppressed by an insupportable sense of loneliness. Unlike Elias Johns, though, he recovers and returns to work 'more calmly', albeit in the face of Margaret's entreaties for him not to proceed with what she sees as a 'dangerous' and 'unhallowed' experiment.[120] The paradigm of the presumptuous electrical experimenter is evident again, as when the dying Elias Johns was described in 'Reminiscences' as a 'vain and presumptuous mortal', 'proud and blasphemous', and the narrator asked 'how fearfully has thy deep sin been visited upon thee, poor child of clay! … Where are thy theories now, thy scoffs and arguings, that led away many a weak spirit into eternal ruin?'[121] The obsession with understanding electricity is portrayed as contravening both the natural and Christian boundaries of man's knowledge, as well as his control of his free will.

Exchanges of scientific practice between electricity and botany are evident in Melchior's nurturing of the mysterious plant and he confesses that he 'loved it like a human thing'.[122] The depiction seems designed to appeal to the era's 'countless amateur botanists', to whom Amy King refers in her study of nineteenth-century botany, as well as to electrical hobbyists.[123] Melchior applies powerful batteries to the 'dewy globules' and eventually a living plant emerges, the leaves of which

---

[117]    Anon. (1853).

[118]    Anon. (1853).

[119]    Anon. (1853), 668.

[120]    Anon. (1853), 670.

[121]    Anon. (1842), 136.

[122]    Anon. (1853), 671.

[123]    Amy M. King, *Bloom: The Botanical Vernacular in the English Novel* (Oxford: Oxford University Press, 2003), 17.

are described as being 'of a sickly white hue, almost like dead flesh'.[124] A small bulbous head and a 'deep sullen purple' flower soon develop, like an anemone in shape and colour 'but of a thick and fleshy texture'; however, it is also described as 'strange, uncouth, misshapen' and a 'wizard tree'.[125] It makes manifest Thomas Carlyle's warning in *Signs of the Times* (1829) of how dangerous technology could be for man, sending up 'over his whole life and activity, innumerable stems, – fruitbearing and poison-bearing'.[126] The tree bears the protean nature of electricity, in adopting human characteristics, for example, when Melchior plants it in the garden and notes its 'unnatural and weird antipathy' towards Margaret.[127] At the same time, the plant reacts like the electric creation it is, responding positively to Melchior's watering but negatively to Margaret's presence, drooping and shrinking from her touch.

The climax of the story seamlessly combines scientific references with visual spectacle, drama and moral allegory, exemplifying the hybridity of writings about electricity. The night before his wedding to Margaret, Melchior dreams of a tree similar to his own, and he is urged by a disembodied voice to eat its fruit. In the morning, he finds his real tree laden with 'gorgeous golden globes ... like pomegranates, of a fiery red' but, when he plucks the fruit, the broken stem swells and turns purple, 'not unlike a tumour on a human body' and a green insect crawls from the 'orifice of the wound'.[128] Melchior describes the insect as 'about the size of a common fly, but snouted and pig shaped, and covered with diminutive bristles'.[129] The depiction resembles closely the electro-crystallisation experiments of Andrew Crosse in 1837, which appeared to result in the production of living mites.[130] Like the fictional Melchior, Crosse was roundly attacked as a charlatan and an atheist, for, as James Secord puts it, it was felt he had 'broken the ancient boundary between life and matter'.[131] Crosse's experiment provoked such a furore that it was widely discussed at all the London institutions and in the popular press, and the issue continued to be debated for decades afterwards.[132] The description of the insects emerging from Melchior's tree corresponds directly with the 'elongated snout', 'elongated snoutish head' and 'bristles' of Andrew Crosse's description, as well as the illustrations of his 'created' mites (see Figure 4.2).[133] Melchior's insects

---

[124]   Anon. (1853), 670.

[125]   Anon. (1853), 671, 672.

[126]   Thomas Carlyle, 'Signs of the Times' (1829), in *The Spirit of the Age: Victorian Essays*, ed. Gertrude Himmelfarb (New Haven: Yale University Press, 2007), 43.

[127]   Anon. (1853), 672.

[128]   Anon. (1853).

[129]   Anon. (1853).

[130]   Andrew Crosse (1784–1855).

[131]   James A. Secord, 'Extraordinary Experiments: Electricity and the Creation of Life in Victorian England', in *The Uses of Experiment: Studies in the Natural Sciences*, ed. D. Gooding, T. Pinch and S. Schaffer (1989), 338.

[132]   Secord (1989), 339.

[133]   Andrew Crosse, 'Electrical Society', *AoE*, 2 (January–June 1838), 355–60.

are 'preternaturally monstrous', 'hideous in form' and 'loathsome in colour';[134] just like the extreme ugliness of Crosse's mites, which prompted suggestions in the press that they should be called *Acarus horridus*.[135]

that they might have originated from the water, and consequently made a close examination of several hundred vessels filled with the same water as that which held in solution the silicate of potassa.  In none of these vessels could I perceive the trace of an insect of that description.  I likewise closely examined the cre-vices and most dusty parts of the room with no better success."

(351) In subsequent experi-ments, this same insect (which it appears is of the genus acarus, but of a species not hitherto observed, and of which a magnified repre-sentation is given in Fig. 148), made its appearance in electrified solutions of *nitrate and sulphate of copper, of sulphates of iron, and sulphate of zinc ;* also on the wires attached to the poles of a battery working in a con-centrated solution of silicate of

Fig. 148.

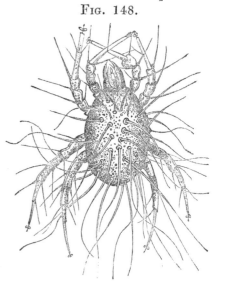

Figure 4.2 *Acarus electricus* (Andrew Crosse, 1837)[136]

When Melchior eats one of the fruits, he describes the sensations that follow as similar to 'the effects of opium'. The effects are so pleasurable that, on their wedding night, he urges Margaret to feast on the fruit too, after which they consummate their marriage and fall asleep. Although he wakes in the morning, to his horror, she is not only dead but covered all over with 'loathsome purple spots' and 'from every ulcerous wound were crawling forth hideous, green, mis-shapen,

---

[134]     Anon. (1853), 673.

[135]     Andrew Crosse, 'Electrical Society', *Literary Gazette*, 1092 (23 December 1837), 818.

[136]     Source: Henry Minchin Noad, *Lectures on Electricity: Comprising Galvanism, Magnetism, Electro-Magnetism, Magneto- and Thermo-Electricity* (London: George Knight and Sons, 1844), 217.

insect reptiles'.[137] Like Charles Hippsley's 'upas', Melchior's tree has turned out to be poisonous. On one level, the tale's ghastly reminder of mortality imitates man's Fall from the Garden of Eden, while, on another, it creates vivid associations between nineteenth-century colonial exploration, sexuality, and diseases thought to be imported from the Empire, such as syphilis, the primary symptom of which is distinctive lesions.[138] The opiate effects of the fruit resonate with Melchior's earlier explanation of his prematurely aged appearance, which he compares to that of 'Eastern dreamers who have fed on opium, and grown unnaturally old before their time'.[139] In a reversal of electricity's purportedly health-enhancing properties, the electrically bred fruit of the plant has sapped Melchior's vitality, youth and energy, enacting instead the laws of magnetism, where like rejects like.

The scenario engages with the politics of the botanical industry and trade, the expansion of Empire and fears about its consequences. In 1853, the same year 'The Tree of Knowledge' was published, the Scottish botanist and traveller Robert Fortune (1812–1880) published his two-volume account of his travels in China for the East India Company.[140] The 'Tree of Knowledge' was published at a mid-point between the Anglo-Chinese Opium Wars, when China sought to restrict British opium traffickers.[141] Melchior's plant, with its bulbous head covered in 'dewy globules' is remarkably like the opium poppy bulb, while its 'sullen purple flower' clearly resembles the plant's subsequent blossom (see Figure 4.3).

So, one might ask at this stage, how are such varied topics as anatomy and insects, trees and opium, related to electricity? What the fictions illustrate is the extent to which electricity was connected to much wider concerns, including the development of other contemporary sciences, colonial trade and politics. Clearly, electricity was in no way distinct from other spheres of Victorian experience, imagination and culture; rather than belonging to a solely scientific sphere, electricity was inextricably linked to numerous other, sometimes unexpected, aspects of contemporary life.

A more specific correlation exists between the 'gorgeous golden globes' depicted in 'The Tree of Knowledge' and the bulbous shape of the electric light bulb – a correspondence which may not be as anachronistic as it might at first appear. Before the 1880s, electric lighting was available predominantly in the form of arc lamps or incandescent gas mantles, and incandescent filament light

---

[137]    Anon. (1853), 674.

[138]    David Schlossberg, *Infections of Leisure*, 3rd edn (Washington, DC: ASM Press, 2004), 335. See also, Stella Pratt-Smith, 'The Other Serpents: Deviance and Contagion in Arthur Conan Doyle's The Speckled Band', *Victorian Newsletter* (Spring 2008: 113), 54–66.

[139]    Anon. (1853), 664.

[140]    Robert Fortune, *Two Visits to the Tea Countries of China and the British Tea Plantations in the Himalaya: With a Narrative of Adventures, and a Full Description of the Culture of the Tea Plant, the Agriculture, Horticulture, and Botany of China*, vols 1 and 2 (London: John Murray, 1853).

[141]    The First Opium War took place 1839–1842 and the Second, 1856–1860.

bulbs were neither efficient nor economical enough to be in widespread use. More recently, though, scholars have suggested that by the 1840s and 1850s there had already been 'considerable progress' in the development of incandescent lamps.[142]

Figure 4.3 Opium poppy flower *papaver somniferum*[143]

In 1850, Joseph Swan was working on the electric light bulb using carbonised paper filaments, and, in 1856, nitrogen-filled (filament) bulbs were used to light up the St Petersburg harbour.[144] Unfortunately, ascertaining with any certainty how much the author of the 'The Tree of Knowledge' knew about early electrical

---

[142]    William J. Hausman, Peter Hertner and Mira Wilkins, *Global Electrification: Multinational Enterprise and International Finance in the History of Light and Power, 1878–2007* (Cambridge: Cambridge University Press, 2008), 11.

[143]    Opium Poppy Flower (*Papaver Somniferum* Midnight). Photograph provided and reproduced with the kind permission of Ray Brown at Plant World Seeds, St. Marychurch Road, Newton Abbot, Devon, TQ12 4SE, UK.

[144]    Hausman et al. (2008). The first incandescent electric light was made in 1800 by Humphry Davy but, in 1878, Sir Joseph Wilson Swan (1828–1914), British physicist and chemist, received the British patent for the carbon filament lamp.

devices is stymied by his/her anonymity. The image of the bulb is most likely to have been adopted by both electrical science and literature from contemporary discussions of vegetation and perhaps other forms of lighting, such as the Polish invention of a new safe, odourless oil lamp in 1853.[145] However, the possibility of a connection represents an interesting dynamic between the terminology used to describe 'natural' items, such as bulbs, and nineteenth-century technological devices. The story was published in the period when electrical devices like bulbs were being devised and so, whether it replicated them, provided inspiration for them, or simply shared a developing idea, perhaps the correspondence should not be entirely ruled out.

In both 'Reminiscences' and 'The Tree of Knowledge', there is a persistent connection between electricity and what is natural or unnatural. The relationship between the three is one of contrast, rather than affinity, which draws on Romantic warnings that 'our meddling intellect/Misshapes the beauteous forms of things'.[146] As Carolyn Marvin points out, the existence of similar oppositions are often evident in 'expert tales' about electricity, which present particular ambivalence about connections between man, Nature and technology. Marvin's description of the relationship is particularly relevant to 'Reminiscences' and 'The Tree of Knowledge':

> Man with his tools knew better than nature; he had foiled and humbled it, laughed at its ignorance, and made it run to do his bidding ... the underlying fabulous and mythic encounter between nature and the magic wand of human invention – a melodramatic, magical confrontation full of the stuff of fairy tales – was disguised in the vocabulary of scientific knowledge and achievement ... Nature damaged or threatened by electricity was evidence of human triumph over instinctual forces.[147]

In both the stories, electricity acts as the opposite of natural processes, such as birth and death, and its productions, such as plants and trees. However, if electricity is evidence of 'human triumph', Elias Johns's and Melchior's victories are short-lived and hollow, with the first being killed by his obsession with electricity and the second barely surviving. Electricity is depicted as a distorting power which, as well as resulting in the protagonists' monomania, reverses the death of a father and causes the death of a son in 'Reminiscences', and gives birth to monstrous insects and slays an innocent bride in 'The Tree of Knowledge'. Overall, writers respond to electricity as a deviant force that had an unnatural influence upon what was otherwise natural. Electricity was perceived to occupy an anomalous position

---

[145]    Alison Fleig Frank, *Oil Empire: Visions of Prosperity in Austrian Galicia* (Cambridge, MA: Harvard University Press, 2005), 56.

[146]    William Wordsworth, 'The Tables Turned', ll.26–7, in William Wordsworth and Samuel Taylor Coleridge, *Lyrical Ballads* (London: J. and A. Arch, 1798), 188.

[147]    Marvin (1988), 116.

within the order of nature; despite being a natural phenomenon, it had destructive qualities that could prey on man's vulnerable sensibilities. In contrast to the perception by scientists and practitioners of electricity as a useful and positive resource, its exploration was portrayed in fiction as one of the period's more dangerously deluded contemporary fascinations.

### The Romantic Supernatural and Modern Futures

Short fictions like 'Reminiscences' bring into question distinctions between earlier forms of 'Romantic' science and nineteenth-century movements towards systematic science between the 1830s and 1880s. The narrator reminisces about a pre-electric time, a time when 'romance ... used to hang about chemistry, physiology, electricity and the rest'.[148] Although the term 'scientist' was coined at the beginning of the period, in 1833 by William Whewell, the nature of what was 'scientific' was still indistinct. As Louise Henson explains, 'apparitions and spectral illusions were widely discussed in the early and mid-nineteenth-century mental philosophy' and particularly so 'in relation to the involuntary functions of the mind, including dreaming, somnambulism, reverie, and more serious cases of mental derangement'.[149] Nineteenth-century fiction authors appear to be keenly aware of the tension between the emerging authority of laboratory-based science and the anecdotal nature of the supernatural, and they explore the fictional possibilities of both. The parameters of the scientific had yet to be decided in the nineteenth century. Investigations of telepathy, telekinesis, mesmerism and the like, as aspects of science, were widespread and, in fiction and non-fiction, scientific advances often contributed to perceived alliances between the supernatural and electricity.

In the short fictions already discussed, Charles Hippsley described himself as a 'man-monster'; Elias Johns's eyes had a 'mystic sort of an expression'; the results of Melchior's electrical experiments were 'almost magical'; and the 100-year-old corpse of Goldsmith was found lying 'round and smooth and life-like'.[150] Each story displays elements of the 'natural' straying beyond what is 'normal'. Viewed as a whole, the instances might appear distant from the concerns of nineteenth-century electricity and science if, as one contemporary reviewer put it, 'supernatural and natural modes of enquiry ... mutually exclude each other'.[151] However, as Fred

---

[148]    Anon. (1842), 123.

[149]    Louise Henson, '"In the Natural Course of Physical Things": Ghosts and Science in Charles Dickens's *All the Year Round*', in *Culture and Science in the Nineteenth-Century Media*, ed. Louise Henson, Geoffrey Cantor, Gowan Dawson, Richard Noakes, Sally Shuttleworth and Jonathan R. Topham (Aldershot: Ashgate, 2004), 115.

[150]    'Mr. Hippsley', 379; 'Reminiscences', 123; 'The Tree of Knowledge', 668; 'An Anachronism', 153.

[151]    Review: 'The Physiology of Common Life', *The Crayon*, 6:3 (March 1859), 71–3.

Botting suggests, science and art often draw on shared cultural references, and science is often 'implicated in and affected by' the 'infectious myth-making' of art.[152] Botting attributes negative associations with electricity to Mary Shelley's *Frankenstein*, in that 'linked to a monster, electricity came to signify the unnatural birth of a dangerous and uncontrollable phenomenon'.[153] In the fictions about electricity so far, portrayals of electrical experimentation can be linked to distorted versions of birth and rebirth not least because, as Carolyn Williams points out, 'of all the resuscitative means, electricity was the most exciting', as well as the most controversial.[154] Nineteenth-century associations between electricity and the unnatural do not, however, relate solely to Mary Shelley's tale or the depiction of monstrosity. Distinctions between what was man-made and what was artificial were articulated well before 1818 and they were often tenaciously linked to electricity. After the eighteenth century, forms of electricity beyond the atmospheric were termed 'artificial electricity'.[155] With the increasing association of experimentation and electricity between the 1830s and 1880s, it was the combination of the two that led to distortions of the natural, rather than electricity alone. Experimentation with electricity was an activity that questioned the status quo and the nature of normality because it meant dealing with the unknown, as a purpose of specialist science. Before mid-century, science itself had yet to be properly defined as an endeavour, so electricity was not simply a matter of science. As a result, literary responses to electricity, whether by scientists or fiction authors, are also often responses to uncertainty and hypothesis.

Botting suggests that fears about electricity were often based on its associations with the new and rational, the secular and humanist.[156] However, the prevailing features of the story to which we now turn seem markedly different. The anonymous short story 'Doctor Beroni's Secret' (1884) was published shortly after 1880, but its late authorship provides an interesting contribution and comparison. The tale opens with the narrator's childhood recollection of gazing at an electrical machine in his father's laboratory, 'half in terror, half in delight, at the fuming and

---

[152] Fred Botting, 'Metaphors and Monsters', *Journal for Cultural Research*, 7:4 (2003), 341.

[153] Fred Botting, *Making Monstrous: Frankenstein, Criticism, Theory* (Manchester: Manchester University Press, 1991), 190.

[154] Carolyn Williams, '"Inhumanly brought back to life and misery": Mary Wollstonecraft, Frankenstein, and the Royal Humane Society', *Women's Writing, the Elizabethan to Victorian Period* 8:2 (2001), 220.

[155] Michael Brian Schiffer, Kacy L. Hollenback and Carrie L. Bell, *Draw the Lightning Down: Benjamin Franklin and Electrical Technology in the Age of Enlightenment* (Berkeley: University of California Press, 2003), 272, 7n. See, for example, Alessandro Volta, *Of the Method of Rendering Very Sensible the Weakest Natural or Artificial Electricity* (London: J. Nichols, 1782) and Giambatista Beccaria, *A Treatise upon Artificial Electricity* (London, 1780).

[156] Botting (1991), 190.

hissing retort'.[157] Immediately electricity is associated with excitement, danger and experiment. The narrator's name is John Glendinning and he tells of how, rather than going to Oxford to study, his father taught him 'chemistry, electricity, and one or two other sciences', before sending him to Germany to be taught by 'the blue-spectacled professors there'.[158] Electricity is characterised as a distinct scientific subject but, by going to Germany rather than Oxford to study medicine, the narrator is shown to have stepped outside ordinary educational conventions and cultural mores, as a maverick personality. As a response to electricity, the narrative is quickly dominated by elements of the supernatural, magic and fairytale, supporting the *Illustrated London News* claim in 1853 that 'industrial technology was unable to dispel the rising tide of spiritualism'.[159] Recent scholarship on the supernatural in nineteenth-century literature proposes that the 'collapsing of time and distance' by new technologies transforming daily life was itself often felt to be 'uncanny'.[160] Rather than nineteenth-century technologies being opposed to the supernatural, they made the material world seem *more* supernatural, and the mysteriousness of electricity simply added to the impression of a world 'full of invisible, occult forces'.[161] Unlike non-fiction popularisations, which dealt with understandings of electricity, fiction writings on the subject were enhanced by readers' lack of scientific understanding.

The instability in the authority of scientific practices, the occult and marginal sciences is apparent in the depiction of electrical experimentation and experimenters. The integrity and authority of the eponymous Dr Beroni is made uncertain by Glendinning's remark that he 'could not be sure whether Dr. Beroni was a man to be trusted or not', not least because of his 'craze for phrenology'.[162] Beroni's laboratory is in his house, placing Beroni's electrical experiments at the centre of his inner, domestic life. Glendinning is given responsibility for charging chemical solutions with electricity, heating them in a furnace and tabulating the results but he describes Beroni's experiments as 'more electricity than chemistry' and a 'jumble of the two'.[163] When Glendinning joins the doctor and his daughter, Ina, for dinner, the systematic and factual nature of scientific practice becomes increasingly hazy as the father and daughter explain their shared fascination for mesmerism, telepathy and card tricks. In a footnote, the author cites a real article on 'Thought Reading', published in the June 1882 edition of the *Nineteenth Century*

---

157    Anon. (1884), 187.

158    Anon. (1884).

159    Quoted in Jason Marc Harris, *Folklore and the Fantastic in Nineteenth-Century British Fiction* (Aldershot: Ashgate, 2008), 10.

160    Nicola Bown, Carolyn Burdett and Pamela Thurschwell, *The Victorian Supernatural* (Cambridge: Cambridge University Press, 2004), 1.

161    Bown et al. (2004).

162    Anon. (1884), 189.

163    Anon. (1884), 206.

magazine, creating an interesting overlap between the interior of the fiction and the external world of its authorship.[164]

Scientific boundaries are brought into question in the story's climax, in which Glendinning reluctantly takes part in a metaphysical experiment with Beroni and his daughter. The three kneel together and stare into a crystal ball, in which Glendinning sees an image of a stranger about to kill Beroni, before suddenly losing consciousness. At this point, the narrator hints at the doctor's involvement with a society called the 'Brothers of the People' and the 'Brotherhood'.[165] The name may refer to the Pre-Raphaelite Brotherhood, founded in 1848 and dominated by the Italian Rossettis; however, *Mafia* organisations originated in the early nineteenth century and one of the largest, the 'Fratellanza' of Girgenti, was legally investigated in 1884, the same year that 'Dr. Beroni's Secret' was published.[166] The conclusion of the story involves a swift return to the focus on electricity when Dr Beroni's previous assistant, Lord Danzil, who has been on the boundaries of the story throughout, is found dead with an electric wire still in his hand, having been killed by 'a current of terrible force'.[167] Meanwhile Dr Beroni is found stabbed, with a note thrust into his jacket declaring him a traitor and, although it is not specified what or whom he has betrayed, it is revealed that the actual aim of Beroni's experiments was to manufacture diamonds.[168]

In 'Doctor Beroni's Secret', the still relatively new practices of electrical science are aligned with interest in mesmerism, phrenology, monomania, and telepathy, all of which are undercut by the final connection with money and industrial manufacturing. The story merges the boundaries between electricity and less explainable realms, such as the supernatural and illusions, magic and fairytale. Beroni performs the role of the exotic and mysterious wizard-magician, while Glendinning is the charismatic suitor for the hand of his beautiful daughter, Ina. Features of melodrama, magic and fairytale have long been recognised in writings about electricity.[169] Neither were they specific to 'popular' fiction; by the second half of the century, as Jack David Zipes points out, 'the use of the fairy tale as commentary was pervasive in high and low culture'.[170] In *Household Worlds*, Dickens refers to readers' 'delight in the Inscrutable' and suggests that man is

---

[164]  Anon. (1884), 194; W.F. Barrett, Edmund Gurney, and Frederic W.H. Myers, 'Thought Reading', *Nineteenth Century*, 11:64 (June 1882), 890–901.

[165]  Anon. (1884), 354, 359.

[166]  Letizia Paoli, *Mafia Brotherhoods: Organized Crime, Italian Style* (Oxford: Oxford University Press, 2003), 40.

[167]  Anon. (1884), 358.

[168]  We might note further also that, while graphite was used as an electrical conductor in the electrodes of arc lamps, diamonds are carbon allotropes, or structurally similar; see Danail Bonchev and D.H. Rouvray, *Chemical Topology: Introduction and Fundamentals* (Amsterdam: Gordon and Breach, 1999), 85–6.

[169]  Marvin (1988), 116.

[170]  Jack David Zipes, *When Dreams Came True: Classical Fairy Tales and their Tradition*, 2nd edn (New York: Routledge, 2007), 159.

always most interested in what is least explainable; indeed, he places that interest at the very centre of electrical science by asking 'would Faraday wield such a magician's rod over the British Institution, if he did not refer a million marvels of nature's forces to the one infinite, incomprehensible power of electricity?'[171] A key appeal of electricity to readers, whether they were looking either for entertainment or scientific understanding, was its danger, its unpredictability and its mystery.

Martin Willis proposes that, 'in the early nineteenth century, to label the form of knowledge "scientific" was to see it as distinct from the magic and occultism of the preceding centuries', while 'to label knowledge "magical" was to suggest it was archaic and without relevance to the contemporary world'.[172] In 'Doctor Beroni's Secret', it is the combination of the scientific and magical that makes it appealing – a combination so attractive that it spawned the creation of several nineteenth-century occult societies.[173] In the stories considered here, the young heroes are repeatedly training to be scientists and it is precisely because they are rational, scientific heroes that their seduction by supposedly outdated and ostensibly less scientific methodologies is so interesting.

In fiction, at any rate, electricity was a part of both new developments of science and older, pre-Christian forms of magic. Glendinning's willingness to experiment beyond the boundaries of science is not a movement away from electricity; instead, it indicates his acceptance of what is portrayed as the true nature of electrical experimentation. As Linda Simon suggests in her discussion of the late nineteenth-century electrical fiction 'The Electric Lady', the hero masters the heroine because he accepts 'electricity and all that it implies: urbanization, new stimuli, and sexual excitement'.[174] The séance room in Beroni's house has been set up in elaborate Greco-Roman style as a pagan temple, with a ceiling of 'deep blue flecked with golden stars', 'statues of white marble' round the walls, 'tripods holding braziers, from which rose clouds of incense, white, blue and rose coloured' and 'weird, sad and strange' music.[175] Beroni remarks on the 'stage effect', saying that the 'Temple of Solomon had the same' because 'all the old religions' resort to similarly enigmatic means.[176] Ancient astrology is presented as superior to modern science when the doctor derides the 'shallow scepticism' of new sciences, exclaiming: 'On what rocks of accumulated experience has each of these mysterious rules and laws been based? And because it is not a "subject"

---

[171]    Charles Dickens, 'A Physician's Ghosts', *All the Year Round*, 15 (August 1859), 346.

[172]    Willis (2006), 34.

[173]    Mystic societies founded in the nineteenth century included the Metaphysical Society, the Hermetic Order of the Golden Dawn, the Order Temporale Orientalis (O.T.O.), the Rosicrucians or the Fellowship of the Rosy Cross, and the Theosophical Society.

[174]    Simon (2005), 164.

[175]    Anon. (1884), 199.

[176]    Anon. (1884), 201.

in competitive examinations, you think it is not a branch of knowledge!'[177] Rather than older practices being separate from modern science, for Beroni, they provide its true foundations. By participating in a ritual within this setting, Glendinning demonstrates his compliance with alternative scientific paradigms.

Scientific heroes begin by investigating science in the short stories considered here, only to find themselves seduced by romance and mythology. A struggle emerges, as Tabitha Sparks suggests, 'between the growing authority of professional medicine and the sentimental and intuitive feelings that inflect the convention of romance'.[178] As if to mirror internal changes, it is when Glendinning shows himself willing to engage with elements beyond the laboratory that the interior décor of the Beroni house alters. The possibility of male–female romance is an integral part of scientific allure; initially, Glendinning refuses to participate in Beroni's séance but, when his daughter Ina asks, he agrees. As scholars acknowledge, 'short fiction has always lent itself readily to what can most accurately be described as the love story' and a 'burgeoning "Romance" literature existed' that emphasised the almost 'supernatural wonder associated with electricity'.[179] Romantic plots operate concurrently with scientific experimentation and curiosity, so that, in several of the fictions, young protagonists fall in love with the beautiful daughter or sister of another scientist. Central to Beroni's power of persuasion is his daughter, Ina, whom Glendinning describes as 'so beautiful, so solitary, so charming in her manner'.[180] From the outset, she and her beauty have a mesmerising and seductive effect upon him. Before Glendinning and Ina, Melchior falls in love with Margaret, Professor Inkleman's daughter in 'The Tree of Knowledge'; the narrator of 'Reminiscences' becomes romantically involved with Elias Johns's sister Catherine; and in 'Mr. Hippsley', Charles pines for another Catherine, whose guardian is his physician. The fictional women are inseparable from the characters, practices and allure of science, and they are as beautiful, unavailable and pure as their names suggest.[181] Their beauty is always simple, unadorned and 'natural', yet there is also something other-worldly about them. Like the phenomenon of electricity, they have the vibrancy of the natural, alongside its dangerous potential. Ina Beroni has 'a white face, and a crown of jet-black hair', 'perfect features' and hands that are 'thin and delicate', while Margaret Inkleman has 'a face radiant with pure and delicate health', as well as a 'warm and witching bloom'.[182] They live with their fathers

---

[177]   Anon. (1884), 203.

[178]   Tabitha Sparks, *The Doctor in the Victorian Novel: Family Practices* (Farnham: Ashgate, 2009), 25.

[179]   Devine Jump (1998), 5; Graeme Gooday, *Domesticating Electricity: Technology, Uncertainty and Gender, 1880–1914* (London: Pickering and Chatto, 2008), 58.

[180]   Anon. (1884), 204.

[181]   The names 'Catherine' and 'Margaret' are Greek in origin and mean 'pure' and 'pearl' respectively; the name 'Ina' has no specific meaning but is a diminutive for various feminine names.

[182]   Anon. (1884), 192, 193; Anon. (1853), 665.

in isolated locations, situations that add to the sense of their being untouched and untainted. Ina dresses in white throughout and is described as 'a recluse, as much out of the world as if she were in a convent', with the purity of a votive maiden set apart from earthly realities.[183]

Ina Beroni's beauty emanates particularly from sharp contrasts of dark and light, in a way that seems to reflect the post-1881 authorship of the story and the introduction of electric lighting. As Carolyn Marvin suggests, electric lights 'invoked the world of classical culture, clean, spare, and geometrically pure'.[184] In the same way, the stark purity of electric light seems to shape the characterisation of women in short electrical fictions. Rather than adding realism, however, it was a change that enhanced the tale's surrealism; electric lights made things appear not to belong 'to the prosaic order of things' but, rather, 'to a natural and supernatural world that seemed nearly, but not quite, beyond man's creative power'.[185] The clean, pure ethereality of Ina and Margaret's beauty is effectively and metaphorically the new form of beauty, which is both electrified and electrifying.

In all the best fairy tales, young heroes fall in love with women who are beautiful and pure. However, in short stories about electricity, the significance of their beauty is more than a useful feature of romantic plots; instead, it establishes an important link between narratives of the fairy tale past and the future of science. The purity and health described in the female characters engages with the aesthetic discourses of the second half of the nineteenth century, which Angélique Richardson describes as 'equating the ugly with disease and beautiful with health'.[186] Electrical investigation may have developed from a study of natural phenomena and a foundation of older philosophies but, in the last quarter of the nineteenth century, it was also the science of modernity and the path to the future. As Richardson explains, the later emergence of the Eugenicist movement sought to portray what Francis Galton termed 'the most appropriate parents of the next generation' in a 'favourable light'.[187] Genetically, the purest and most beautiful women were the most appropriate partners for science and the scientists who were creating the future.

Responses to electricity in fiction are never divorced from science. They engage with electricity as a distinct entity but they also treat it as inseparable from both past and contemporary issues. Electricity is both the science of the future and a vehicle for the past. As James Mussell suggests, 'electricity can stand metonymically for the modern age' in that it makes distinct 'the lived experience of space and time'

---

[183]    Anon. (1884), 204.

[184]    Marvin (1988), 165.

[185]    Marvin (1988).

[186]    Angélique Richardson, *Love and Eugenics in the Late Nineteenth Century: Rational Reproduction and the New Woman* (Oxford: Oxford University Press, 2003), 80.

[187]    Richardson (2003). Francis Galton, Galton Papers, 138/4, 10. The Eugenics Education Society was founded 1908, in which Sir Francis Galton FRS (1822–1911), Karl Pearson FRS (1857–1936) and Major Leonard Darwin (1850–1943) were instrumental.

of the past and present.[188] The reluctance to relinquish the past adds a dystopian tone to the fictions; whatever progress electricity might have offered, it was almost always perceived to have a darker side. Electricity provided ideal material for nineteenth-century short-fiction authors, for it retained the mystery of man's older, natural origins but also provided an icon of nineteenth-century 'rational' science and the 'real-life' technologies that were perceived to be shaping the future.

---

[188]   Mussell (2007), 189.

# Chapter 5
# Electrical Utopias: *Another World*

Fiction writings about electricity are located in more distant regions in *Another World, or Fragments from the Star City of Montalluyah* (1873) by Benjamin Lumley.[1] I approach the novel as a case study because it demands to be treated differently from the writings already examined. The novel's basis as a fiction is relatively unusual and contemporary responses to it provide several rare insights into contemporary perceptions of writings about electricity. By examining contemporary reviews, we can trace more accurately the interactions between the text and its readerships, and their influence in the novel's transition from popularity to obscurity. The reviews also allow us to explore further questions of literary and scientific authority, about how far popular interest in a text depends on the accuracy of information, why authors might choose to deliberately distort that accuracy, and whether, in an era of rapid scientific and literary change, the otherwise fixed nature of scientific information becomes more productively mutable, open to interpretation and creative transformation, both by writers and scientists. In these aspects, as well as several others, *Another World* is a particularly fruitful resource.

Benjamin Lumley (1811–1875) was a solicitor who, from 1841, became opera manager of the illustrious His Majesty's Theatre, London.[2] His pen-name, however, was 'Hermes', a figure who, significantly, was the god of transitions and boundaries, moved as fast as lightning, and is often referred to as the conductor of the dead to the afterlife.[3] Between 1851 and 1855, Lumley had been a science journalist for the *Critic* magazine, reporting on contemporary developments in a huge range of sciences from electricity to anatomy, in two series entitled 'Scientific Summary'

---

[1]   Benjamin Lumley, *Another World, or, Fragments from the Star City of Montalluyah, by 'Hermes'* (London, 1873). Lumley published his first novel at the age of 51, a relatively inconsequential moral romance called *Sirenia: Or, Recollections of a Past Existence* (1862); see 'Sirenia', *Saturday Review*, 13:336 (5 April 1862), 390. Lumley's other writings include a standard book on *Parliamentary Practice on Passing Private Bills* (1838); an account of a legal dispute with the Earl of Dudley, *The Earl of Dudley, Mr. Lumley and Her Majesty's Theatre* (1863); and his memoirs, *Reminiscences of the Opera* (1864).

[2]   Lumley was born Benjamin Levy, the son of a Canadian merchant, and he was schooled in Birmingham. L.M. Middleton, 'Lumley, Benjamin (1811/12–1875)', rev. John Rosselli, *Oxford Dictionary of National Biography* (Oxford University Press, 2004; online edn, May 2007) [http://ezproxy.ouls.ox.ac.uk:2117/view/article/17174; accessed 9 December 2010].

[3]   Everett F. Bleiler, *Science-Fiction, the Early Years* (Kent, OH: Kent State University Press, 1990), 359.

and 'Science and Inventions'.[4] Knowing this background is vital in reading *Another World* because, unlike most of the fiction writers discussed so far in this book, it indicates that Lumley was relatively well-informed about the science of electricity. The fictional 'electricities' of his novel differ wildly from actual scientific methods but not apparently as a result of his ignorance. In the 1850s, he reported on the electrical researches of François Arago, David Brewster, John Tyndall and Robert Hunt, to name but a few, and he was a stickler for accuracy; he postponed summarising one of Faraday's papers, for example, due to 'correctness in scientific reports being of much more importance than mere rapidity of communication'.[5] As a literary response to electricity, *Another World* appears to be an extravagant fantasy about electricity as a perfect route to technological advancement and social progress. The articulation of ideas about electricity is necessarily oblique, as indicated in the earlier discussion of engagements by Faraday and Maxwell, making it futile and even misguided to seek exact parallels. As Gillian Beer points out, 'we should not look for stable one-to-one correspondences between scientific exposition and literary creation'.[6] While correspondences can be identified in writings about electricity, multi-faceted perceptions of the phenomenon also act as sources of instability. *Another World* occupies the unusual position of being informed by Lumley's background of documenting early nineteenth-century developments of electrical science, but it can be approached as a form of 'science fiction', in that it 'compels the reader to revise presumptions of plausibility'.[7] Despite the novel's apparent whimsicality, Lumley was vehemently opposed to those he described as 'the charlatan and the quack'; indeed, he claimed that 'no approbation, however stern and unqualified, can be too severe' for what he described as their 'pseudo-science'.[8] He felt it his duty as a reporter to point out 'the errors and follies of the day'; however, he also thought it perfectly acceptable to indulge in what he termed 'the vagaries of the imagination, but all in good faith'.[9] Lumley's use of the term

---

[4]    'Hermes', 'Summary of Science', *Critic*, 10:257 (15 December 1851), 609; 'Scientific Summary', *Critic*, 14:336 (2 April 1855), 170; 14:340 (1 June 1855), 264; 14:341 (15 June 1855), 288; 14:348 (1 October 1855), 488; and 'Science and Inventions', *Critic*, 11:258 (1 January 1852) to 14:347 (15 September 1855).

[5]    Hermes, 'Summary of Science', *Critic*, 10:257 (15 December 1851), 609. The paper Faraday read was 'On the Lines of Magnetic Form, their Disposition within a Magnet and through Space' at the Royal Society on 11 December 1851.

[6]    Gillian Beer, *Open Fields: Science in Cultural Encounter*, 2nd edn (Oxford: Oxford University Press, 2006), 168.

[7]    David Seed, ed., *A Companion to Science Fiction* (Blackwell Publishing, 2005).

[8]    Hermes, 'Science and Inventions', *Critic*, 11:261 (16 February 1852), 102. Lumley was objecting to spurious explanations of what was termed 'Vital Magnetism'. His use of the term 'approbation' meaning 'approval' may seem mistaken in this context; however, Lumley appears to employ the term's original meaning, now obsolete, of 'probation' or 'trial'. See, for example, Shakespeare *Measure for Measure* (1623) i. ii. 166: 'This day, my sister should the Cloyster enter, And there receive her approbation' (OED).

[9]    Hermes (1852).

'fragments' is apposite in echoing the title of John Tyndall's volume, published just two years earlier, *Fragments of Science for Unscientific People* (1871). *Another World* is not an attempt to represent science faithfully or accurately; instead, it is a work of imagination that responds to and adapts developments of electricity, in order to engage with the further hopes and concerns of the period.

The novel is set on a fictional planet called Montalluyah, in an unspecified time when electricity is the basis for almost every aspect of life. Electricity and light are considered to be the essence of spirituality, 'the point of union between the immortal soul and the perishable portions of man' – a relationship between light, spirit and body that echoes many of the previously discussed characterisations of electricity as enigmatic and spiritual.[10] To this depiction, Lumley adds ideas about how electricity might affect a range of other contemporary concerns about the future, such as fuel supplies, food production and medical treatments. The reviews of the novel offer evidence of contemporary readers' perceptions; by considering them first, I seek to place the novel's reception on a par with the text, adopting for the treatment of a fiction of science Jonathan Topham's suggestion that

> Until the readers for science are brought into the foreground, and considered as actively engaging with what they read, the processes by which the authoritative audience-relations of science were actually accomplished will not be adequately understood.[11]

By foregrounding reviews, I aim to embrace a wider range of responses to the novel's core themes, and particularly its focus on electricity. The reviews of *Another World* suggest that, as a response to electricity, it was perceived as helpful and instructive to more general 'readers for science'. Today, *Another World* is virtually unheard of and, although it is mentioned in Bleiler's catalogue of early science fiction, it has not been the focus of any further scholarship.[12]

## Reception and Interpretation

When Lumley's novel was published in March 1873, it was heralded as 'an extraordinary and wide-spread sensation' and, by June, a third edition had already been published, 'about which intelligent people in all circles have been talking for months past'.[13] These claims appeared in the *Musical World*, a prestigious weekly

---

[10]   Lumley (1873), 65.

[11]   Jonathan R. Topham, 'Scientific Publishing and the Reading of Science in Nineteenth-Century Britain: A Historiographical Survey and Guide to Sources', *Studies in History and Philosophy of Science*, part A, 31:4 (2000), 563.

[12]   Bleiler (1990), 359.

[13]   'Another World', *Musical World*, 51:14 (5 April 1873), 220; Dishley Peters, 'Another World', *Musical World*, 51:27 (5 July 1873), 451.

for which Lumley regularly wrote.[14] His influence over the magazine's contents should not be underestimated, beyond his role at His Majesty's Theatre.[15] No editor is cited in the magazine's records and, with Lumley's profession unknown at the time, his own possible promotion of the novel should be kept in mind. The volume's pricing might offer an alternative way to assess the novel's popularity but the mid-nineteenth-century introduction of free circulating libraries makes retail pricing less relevant to readership figures or distinctions.[16] More reliable testament to the novel's popularity is that extracts continued to be published not only in *Musical World* but also in Dickens's magazine *All the Year Round*, for over a year after the publication of the volume's first edition.[17] The extracts appeared under the relevant topic from the novel or in the series 'Planetary Life', many of them the same extracts, appearing in *All the Year Round*, first, and then in *Musical World* some two weeks later.[18] In terms of material form, the 'branding' of extracts in *All the Year Round* was not as apt; they were printed with a bamboo-style border while in *Musical World* they were always printed with wavy border lines, reminiscent perhaps of electrical waves (see Figure 5.1).

The novel was published in volume form under the 'Hermes' pseudonym, keeping the author's identity a mystery in much the same way as Bulwer-Lytton's anonymous publication of *The Coming Race*. Any degree of public credibility the novel might have drawn from Lumley's previous scientific reporting would have been negligible by this point anyway, after a two-decade gap; but the anonymity performed another function, as the ground for the novel's simultaneously mediated and first-person narration. The author claims that he is actually just the novel's editor and the story has been dictated to him by the ruler of Montalluyah. Such a far-fetched explanation meets with inevitable derision from modern readers

---

[14]    *The Musical World: A Weekly Record of Musical Science, Literature, and Intelligence*, 1:1 (18 March 1836) to 71:4 (24 January 1891).

[15]    His Majesty's Theatre was rebuilt in 1869, after being destroyed by fire, and remained empty until 1874, when it was used for Revivalist spiritual meetings; theatrical performances did not resume until 1877. The theatre now changes its name according to the gender of the British monarch (Her Majesty's Theatre: www.london-theatre-tickets.org.uk/her-majestys-theatre-history.htm; accessed 20 August 2011).

[16]    The first edition was priced at 12 shillings, higher than *The Coming Race*, which cost seven shillings and sixpence; the third edition had an increased price of 15 shillings (see advertisement, *Evening Post*, 10:4 (20 February 1874), 4.

[17]    Hermes, 'Planetary Life', *All the Year Round*, 11:257 (1 November 1873) to 11:265 (27 December 1873) and from 52:11 (14 March 1874) to 52:20 (16 May 1874).

[18]    Extracts cited the source as either the novel or *All the Year Round*; see 'Planetary Life', *All the Year Round*, 11:257 (1 November 1873), 5 and 'Planetary Life', *Musical World*, 51:46 (15 November 1873), 770. On other weeks, there would be an extract in *All the Year Round* and an additional commentary by Hermes in *Musical World*, with a different extract (see 'Planetary Life', *All the Year Round*, 11:260 (22 November 1873), 84 and 'Planetary Life', *Musical World*, 51:47 (22 November 1873), 783).

and prompts Bleiler, for example, to dismiss the novel as 'an infantile effort'.[19] Some contemporary readers were also dismissive, describing it as 'the flight of a wild fancy – the freak of a keen imagination' and demanding that, in order to be believed, 'we want a proof!'[20] Others were more forgiving, though, reading it as gently satirical; as *Macmillan's Magazine* suggests, the author had too much of 'a glow of good humour ... to find room on his lip for a sneer'.[21] Perhaps more surprisingly, the *Sunday Times* judged that the novel 'gives a faithful, though at present incomplete, account of an actually existing world, whose inhabitants are formed like those of our own planet'.[22] The reviewer's comment implies, very over-generously, that some readers might have believed an entire population could 'actually' exist on another planet.

The novel plays on persistent connections between electricity and the nature of physical phenomena – issues not set apart from scientific explorations of electricity. As Peter Harman suggests, they were also 'interwoven with debates on the nature of physical reality'.[23] The author's claims that the novel was dictated by the unearthly figure of Montalluyah's ruler prompted the *Spiritualist*'s reviewer to suggest that 'it is not impossible that spirits from another planet may sometimes communicate, with more or less precision through a medium'.[24] In the *Morning Post*, too, Hermes's views of electrical science were noted as 'peculiar'; electricity was still considered to be both 'the great executive of nature' and 'almost as occult as what is called spiritualism'.[25] The *Saturday Review*, in contrast, described the novel as 'a very curious book', the 'exact purport' of which was hard to understand; the reviewer praised the novel's 'plain, matter of fact, and even minute descriptions' as entirely different from the 'spiritual manifestations which are fast taking place among the bores of the period' in 'this table-rapping age'.[26] The three reviews capture the debates over 'Spiritism' that dominated the context of the novel's publication and other literary responses to electricity. It is worth noting too that, in 1869, not long before the novel's publication, the London Dialectical Society was formed to report on spiritual manifestations of unexplained phenomena, with which electricity was frequently associated. As the *Orchestra* magazine reported, the Society invited several leading scientists to participate but G.H. Lewes refused

---

[19]    Bleiler (1990), 360.

[20]    'Another World (from Lloyd's Weekly Newspaper)', *Musical World*, 51:11 (15 March 1873), 159.

[21]    'Another World', *Macmillan's Magazine*, 28:164 (June 1873), 140.

[22]    'Music in Another World (from The Sunday Times)', *Musical World*, 51:15 (12 April 1873), 227.

[23]    Peter Harman, *Energy, Force and Matter: The Conceptual Development of Nineteenth-Century Physics* (Cambridge: Cambridge University Press, 1982), 9.

[24]    'Another World (from The Spiritualist, March 15)', *Musical World*, 51:17 (26 April 1873), 263.

[25]    'Another World (from the Morning Post)', *Musical World*, 51:14 (5 April 1873), 211.

[26]    'Another World', *Saturday Review*, 35:900 (25 January 1873), 123.

## PLANETARY LIFE.

### BY HERMES.

### NO. IV. FLOWERS AND BIRDS IN MONTALLUYAH.

There are more things in heaven and earth, Horatio,
Than are dreamt of in your philosophy.

To the delicately organised inhabitants of Montalluyah the flower-garden is an important source of enjoyment, and an object of constant solicitude ; for we have difficulties in horticulture which are to you comparatively unknown, while, on the other hand, we obtain results to which you can offer nothing in comparison.

undoubtedly rapid, and this rapidity, which, when excessive, causes a diminution of strength, is increased by the use of the artificial rain. A counter-agent is therefore required, and we were fortunate in discovering a liquid which, applied to the extremities of a plant, checks the progress of its growth, without doing any injury. We also found that the same liquid, when rubbed on the wrists and veins of the human body, had the power of retarding the circulation—a quality, in our climate, extremely valuable.

The liquid is extracted from the "Vesual"

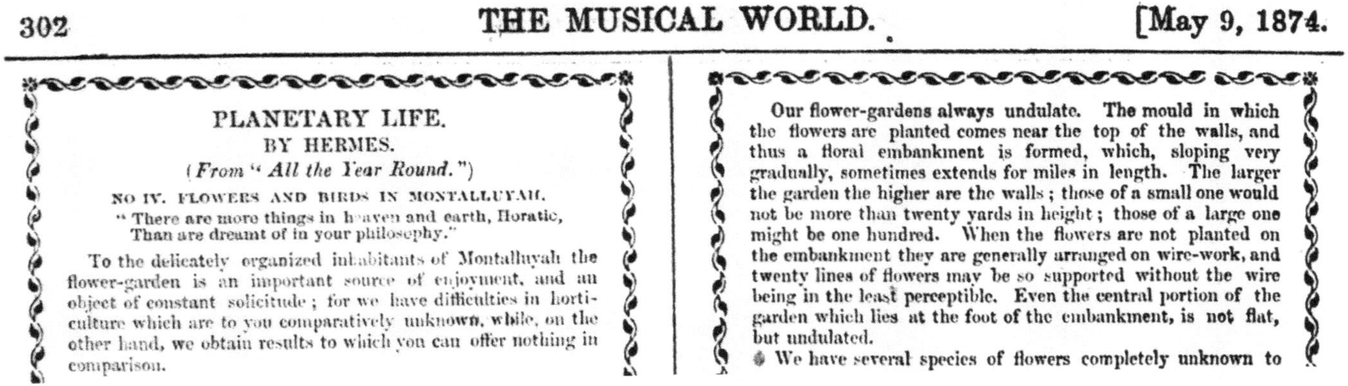

# THE MUSICAL WORLD.

## PLANETARY LIFE.
### BY HERMES.
*(From "All the Year Round.")*

No IV. FLOWERS AND BIRDS IN MONTALLUYAH.

"There are more things in heaven and earth, Horatio,
Than are dreamt of in your philosophy."

To the delicately organized inhabitants of Montalluyah the flower-garden is an important source of enjoyment, and an object of constant solicitude; for we have difficulties in horticulture which are to you comparatively unknown, while, on the other hand, we obtain results to which you can offer nothing in comparison.

Our flower-gardens always undulate. The mould in which the flowers are planted comes near the top of the walls, and thus a floral embankment is formed, which, sloping very gradually, sometimes extends for miles in length. The larger the garden the higher are the walls; those of a small one would not be more than twenty yards in height; those of a large one might be one hundred. When the flowers are not planted on the embankment they are generally arranged on wire-work, and twenty lines of flowers may be so supported without the wire being in the least perceptible. Even the central portion of the garden which lies at the foot of the embankment, is not flat, but undulated.

❀ We have several species of flowers completely unknown to

**Figure 5.1 Comparison of borders on extracts of *Another World* (1874)**

*Source:* Hermes, 'Planetary Life', *All the Year Round*, 12:283 (2 May 1874), 53; 'Planetary Life', *Musical World*, 52:19 (9 May 1874), 302.

and T.H. Huxley declared, 'better live a crossing-sweeper than die and be made to talk twaddle by a "medium" hired at a guinea a *séance*'.[27] After two and half years, Dr James Edmunds (the committee's Chair) was forced to admit that apparently spiritual phenomena were of a 'thoroughly contemptible nature' and nothing more than 'unconscious action', 'self-delusion' or 'imposture'.[28] Within this context, the title of *Another World* may well present a satirical association between the novel's extra-terrestrial location and contemporary fascination with the supernatural.

The novel's purpose as a utopian fiction was more widely agreed upon. The society of Lumley's fictitious planet is overtly ideal and electricity is fundamental to its advanced state. There is no suggestion of ambivalence; indeed, the narrator asserts that 'we were greatly aided by our extended knowledge of electricity', which has allowed technological and social progress.[29] The narrator emphasises that, on Montalluyah, elements do not have to operate in the same way as on Earth, thereby creating leeway for the novel to stray from actual contemporary science. Many of Lumley's descriptions appear, at least initially, to refer to the realities of nineteenth-century electrical science. He claims, for example, that the Montalluyahns had recognised that 'all electricities were in reality one and the same', in reference to the unity proposed by Michael Faraday, although this is not specifically mentioned. Perhaps anticipating objections, the narrator praises ideas about unified electricity as arrived at by 'truly great men, for they had opened the gates of science'.[30] The narrator declares that 'the principle' of unified electricity is right, but he then suggests that 'as was subsequently shown, the application and the conclusion were wrong'.[31] Lumley's fictional narrator can depart with impunity from the realities of electrical discovery, forestalling criticism about his own lack of scientific knowledge and allowing, as a result, greater imaginative possibilities of electricity to be explored. He is free to assert, for example, that 'tangible and visible proofs exist beyond doubt that every kind of body and substance, whether animate or inanimate, contains electricity of its own'.[32] Within his fictional framework, he does not need to pause to present the detail of any of those 'proofs'.

Completely in opposition to the propositions of contemporary electrical science, the narrator differentiates its different types, claiming that

> Some electricities are diffused and attenuated; some are concentrated; others are so tenacious of the body to which they belong that they are all but steadfast.

27    'Spiritism and its Raps', *Orchestra*, 17:421 (20 October 1871), 42.
28    'Spiritism and its Raps' (1871).
29    Lumley (1873), 54.
30    Lumley (1873), 55.
31    Lumley (1873).
32    Lumley (1873), 56.

Some are sympathetic; some antipathetic, attracting or repelling each other; some mingle gently; others, when brought into contact, cause violent explosions.[33]

To make his explanations more credible, Lumley adopts concepts and vocabulary from the sciences; his reference to 'diffusion', for example, referred to the early nineteenth-century description of how gases and liquids moved.[34] He also employs terms in relation to electricity that came into common use several years later. 'Attenuation', for example, was an established general synonym for 'thinning' or 'diminishing' but only in the 1880s was it used to refer to the lessening of an electrical signal or current, as Lumley uses it.[35] The example hints at the unseen contributions that literature might have made to scientific terminologies, despite the difficulty of tracing their routes of transmission. Indeed, as the novel illustrates, Lumley shows remarkable prescience in imagining technologies that would not be discovered until almost a century later.

## Electricity, Nature and Production

In *Another World*, the natural world is the source of all electricity: as the narrator explains, 'many beasts, birds, insects, fish, reptiles, trees, plants, water, in short, all substances organic and inorganic, possess each its own peculiar electricity'.[36] Rather than being portrayed as artificial, as electricity has been in the fictions discussed previously, it is presented as a source of fuel inseparable from other products in the natural world, which is simply harnessed by the inhabitants of Montalluyah. The novel describes the social and moral advances on the planet, almost all of which arise from advances in the processing and technology of electricity. The depiction goes beyond simply portraying the results of using electricity, as Bulwer-Lytton did; instead, the novel offers a detailed vision of how electricity might answer some of the most pressing issues of 1870s Britain.

Technological utopianism was a characteristic feature of 1870s literature, which provided the foundation for similar fictions in the 1880s.[37] While *Another World* might be read as simple literary escapism, it also posits suggestions for some of the period's key industrial concerns. While electricity is used partly 'to delight and

---

[33]   Lumley (1873).

[34]   John Dalton, FRS (1766–1844), *New System of Chemical Philosophy*, vol. I (Manchester: Printed by S. Russell, for R. Bickerstaff, London, 1808–1827), 191.

[35]   Lord Rayleigh, *Phil. Mag.*, 5th Ser. XXII. 490 (1886); Heaviside, *Electrician*, 143/2, 24 June 1887 (OED).

[36]   Lumley (1873), 56–7. Ideas that appear to be undergoing something of a scientific revival; see Gwyneth Dickey, 'Stanford Researchers Find Electrical Current Stemming from Plants', *Stanford University News Service* (13 April 2010) [http://news.stanford.edu/pr/2010/pr-electric-current-plants-041310.html; accessed 22 November 2010].

[37]   For example, John Macnie (Ismar Thiusen), *The Diothas, or Looking Forward* (1883), and Edward Bellamy, *Looking Backward* (1888).

instruct the people', to enhance singing and produce soothing music, most of its uses are more sober.[38] The economic climate at the time makes this unsurprising, as industrial growth in Britain registered a sharp fall in the 1860s and 1870s.[39] Modern scholarship suggests that the utopian impulse in the period's literature was allied to the 'Great Depression' between the mid-1870s and the mid-1890s, which exposed a distinct decline in Britain's industrial supremacy.[40] How the country's industrial productivity and future progress would be powered was a central question in what Gowan Dawson describes as the 'ever more "technocratic"' society of the 1870s.[41] It was an especially transitional period in terms of fuel; with diminishing supplies of inefficient wood, the 'phenomenal expansion' of coal mining began in earnest.[42] Electricity was in competition with gas, coal and, from the 1850s, kerosene, as the fuel of the future. Lumley did not have the scientific knowledge to suggest actual fuel alternatives, and *Another World* does not explicitly propose the adoption of Montalluyahn sources or methods. Instead, electricity appears to be portrayed as an exemplar of a clean, efficient and abundant fuel source, and one that might offer the prospect of reliable future progress.

The novel's concerns about fuel related to the wider issue of what we now call 'sustainability', particularly in the novel's proposed use of electricity for agricultural production. In 1871, less than two years before *Another World* was published, the Siege of Paris had resulted in horrific stories of starvation in the British press.[43] British fears about the reliability of food supplies were intensified further by reports of the 1868 and 1870 famines in India, in which almost one and a half million people died of starvation.[44] Yet rather than focusing on the production of food, *Another World* offers an extensive depiction of how electricity might be used to improve flower production. Careful consideration of the novel in relation to contemporary fears, as well as today's hindsight, indicates that the target of Lumley's satire was the frivolity of electrical applications, in the face of much greater concerns of life and death. The irony was lost on most contemporary readers, though; many of whom still thought electricity would cure the world's ills.

---

[38]    Lumley (1873), 57–8, 64.

[39]    M.W. Kirby, *The Decline of British Economic Power Since 1870* (London: Routledge, 2006), 2.

[40]    Matthew Beaumont, 'The Party of Utopia: Utopian Fiction and the Politics of Readership, 1880–1900', in *Exploring the Utopian Impulse: Essays on Utopian Thought and Practice*, ed. Michael J. Griffin and Tom Moylan (Bern: Peter Lang, 2007), 167.

[41]    Gowan Dawson, 'Literature and Science under the Microscope', *Journal of Victorian Culture*, 11:2 (Autumn 2006), 312.

[42]    George Benedict Baldwin, *Beyond Nationalization: The Labor Problems of British Coal* (Cambridge, MA: Harvard University Press, 1955), xix.

[43]    Robert Cedric Binkley, *Realism and Nationalism, 1852–1871* (New York: Harper and Brothers, 1935), 296.

[44]    Richardson Benedict Gill, *The Great Maya Droughts: Water, Life, and Death* (Albuquerque: University of New Mexico Press, 2001), 91.

The artificial light on Montalluyah has a particular, almost feminised beauty – 'not as vivid as that of the external world, but subdued and beautifully soft'.[45] Electricity is the source of organic production, whereby plants 'attract a sufficient quantity of electricity of the ground' and when 'the two electricities are, as it were, married, their united heat and power force the seed to burst', a nostalgic image that calls up Erasmus Darwin's eighteenth-century electrical vision of the 'two electric streams that conspire' at the beginning of creation.[46] The beauty and scents of natural flowers are increased by electricity but it also creates distinctly nineteenth-century productions, whereby metal flowers are created, based on their natural counterparts.[47] There is something distorted and grotesque about the exaggeration of the natural flower and its replication in metal. In a strange combination of nineteenth-century hydroponics and electroplating, the metal is shaped around the original flower and a 'bag of sympathetic electricity' is 'arranged to fit closely around the form of the metal-flower in such a way that the electricity has no escape'.[48] The fragile natural form is trapped and held in place by the circulating electricity, in a metaphor of electricity's effect on what is natural. The narrator explains that, in the process of electroplating:

> It is essential that the charge should be sufficiently strong to modify or overpower the electricity already existing in the plant, in order to change the form which this would otherwise take.[49]

Rather than the Romantic depiction of science revealing nature to expose electricity, electricity becomes the dominant force, overpowering the natural form sufficiently to change it. Like some of the short-fiction portrayals, electricity's distorting effects are evident, as if to expose its underlying falsifying capacity and incompatibility with organic forms.

Electricity from plants produces an expanded colour spectrum on Montalluyah, not just 'roses, pink, blue, green, lilac, brown', but also 'fire-colour' and 'sun-colour'.[50] Lumley's fascination with the colour spectrum reflects contemporary scientific interest in the subject, evident in James Clerk Maxwell's published papers on colour, its perception and colour 'blindness' between 1855 and

---

45  Lumley (1873), 80.
46  Lumley (1873), 134, 132; Erasmus Darwin, 'The Temple of Nature: Canto III. Progress of the Mind' (1802), l.21. The reference to 'electric streams' represents an early version of the metaphor used in *Villette* by Charlotte Brontë and in *Middlemarch* by George Eliot.
47  Lumley (1873), 81.
48  Julius von Sachs (1832–1897), German botanist and physiologist. See Julius Sachs and Sydney Howard Vines, *Text-Book of Botany, Morphological and Physiological*, 2nd edn (Oxford: Clarendon Press, 1882). Hydroponics was originally investigated in the seventeenth century by Francis Bacon and John Woodward.
49  Lumley (1873), 134.
50  Lumley (1873).

1872. Lumley's scientific reporting preceded Maxwell's work in the 1850s, but Maxwell's paper 'On Colour Vision' was published in 1872, the year before the publication of Lumley's novel.[51] The fictional framework of *Another World* allows electricity, light and colour to be bound together, so that the electricities of birds and goldfish give 'lovely colours', 'moss gives a colour resembling fire-sparks' and 'frogs produce a beautiful violet'.[52] Precisely how they do so is not explained but the fiction touches on genuine electrical theories, for example, of popular early nineteenth-century reports of violet as the colour Sir Isaac Newton identified in the short-wavelength end of the visible spectrum.[53] Violet was strongly associated with femininity, partly after it was categorised as such by the German painter and theorist Phillip Otto Runge in 1809.[54] The colour became tremendously fashionable after its introduction in the textile industry, as an aniline dye in 1856.[55] The narrator's reference to violet may be made in passing but it may also have been a conscious effort to capture the interest of female readers.

The production of colour is portrayed as a harmless manipulation of the natural world, as invented as the process of hunting, in which electricity also plays a vital role. Wild birds are caught by using a combination of seeds and 'sympathetic electricity', with a long hollow metal tube. The tube has a globe at the bottom containing a powerful acid, which is released when the bird alights, causing 'a drop of acid in the globe to escape into the tube, and so set in movement a current of electricity'. Being 'calmed by the electricity', the birds do not 'flutter or struggle when thus secured' and netted.[56] Similarly, although vivisection is practised, 'the animal's eyes are bandaged, so that he does not even know what is going on, but is free from pain'.[57] The reassurances were likely aimed at supporters of the growing anti-vivisection movement, which resulted in the law for the Protection of Animals Liable to Vivisection in 1875, shortly after the publication of *Another World*.[58]

The electrical applications Lumley depicts tend to address how activities, such as electroplating flowers, colour production and hunting, are pursued rather than why. The novel can be read as a flawed satire or, alternatively, as a pastiche of flawed understandings of electricity resulting in ill-conceived solutions to

---

[51]    James Maxwell, 'On Colour Vision', *Royal Institution Proceedings*, 6 (1872), 260–271.

[52]    Lumley (1873), 135.

[53]    For example, William Chambers and Robert Chambers, *Chambers's Information for the People*, vol. 2 (Edinburgh: W. and R. Chambers, 1842), 52; David Brewster, *The Life of Sir Isaac Newton* (London: John Murray, 1831), 68.

[54]    John Gage, *Color and Meaning: Art, Science, and Symbolism* (London: Thames & Hudson, 1999), 35.

[55]    Herbert Norris and Oswald Curtis, *Nineteenth-Century Costume and Fashion* (Mineola: Constable, 1998), 134.

[56]    Lumley (1873), 58.

[57]    Lumley (1873), 61.

[58]    Nicolaas A. Rupke, *Vivisection in Historical Perspective*, Wellcome Institute for the History of Medicine (London: Croom Helm, 1987), 263.

contemporary problems. The fictionality of the novel was often understood by readers, even if they were somewhat confused as to the author's intent. On the one hand, the *Morning Post* praised the novel as 'a good, if not sublime, effort of creative ability'.[59] The *Echo*'s reviewer claimed, on the other hand, that 'the mode of action ascribed to [electricity] would make electricians stare' and that the author's knowledge of electricity was 'not hazy enough for jest, yet too hazy for sober earnest'.[60] Even so, he praised the social applications of electrical inventions in the novel and claimed that, for readers who still knew little about electricity, the novel offered 'amusement combined with no little instruction'.[61] The novel's 'no little instruction' could be understood as none at all; however, most of the novel's reviews indicate that it was viewed as an effective and entertaining way of introducing non-specialist readers to concepts of electricity.

## Electrical Futures

Lumley's novel appears to be a satire that missed its mark, one that was understood by some readers to be a genuine introduction to electrical concepts, as the reviews indicate. The novel is precariously positioned between science, satire and utopianism, but, as Peter Godfrey-Smith suggests in his discussion of how model-based scientific practices relate to fictional models, 'the behavior of the imaginary system is explored, and this is used as the basis for an understanding of more complex real-world systems'.[62] *Another World* presented an imaginary existence, by which readers attempted to understand electricity and its future possibilities. Their interest was engaged by a combination of motivations. The fictional framework was free of material, technical and political restrictions, allowing readers not just entertainment but also the possibility of being part of future developments. The book's liminal character exploited an intriguing position that drew on both fiction and science, and yet, as I consider further now, it also occupied a position between them.

Scientific theory does not inevitably precede fiction portrayals or provide the basis for imaginary depictions. The wider non-specialist audience included those developing the whole range of contemporary sciences, many of whom had no more understanding of electricity than other non-specialist readers. Contemporary scientists were no different from other readers, in enjoying reading novels; to take just two examples, James Clerk Maxwell notes in his diary, 'I have been reading

---

[59]  'Another World (from The Morning Post)', *Musical World*, 51:14 (5 April 1873), 211.

[60]  'Another World (from The Echo)', *Musical World*, 51:18 (3 May 1873), 279.

[61]  'Another World' (1873).

[62]  Peter Godfrey-Smith, 'Models and Fictions in Science', *Philosophical Studies*, 143:1, Models, Methods, and Evidence: Topics in the Philosophy of Science. Proceedings of the 38th Oberlin Colloquium in Philosophy, 143 (March 2009), 102.

*Villette* by Currer Bell alias Miss Brontë', and it is known that Charles Darwin's wife, Emma, regularly read novels to him.[63] While scientists' reading of fiction is not always clearly documented, these instances indicate that fiction was read by them. The ideas portrayed in fiction may well have helped inspire scientific efforts; in fact, all the more so, if the ideas were exciting but the methodologies poor.

The lack of methodological rigour in Lumley's novel is undeniable but the novel's key themes are not entirely divorced from the actual science. David Samuelson's scholarship on twentieth-century science fiction argues that the depiction of science in fiction is not dissimilar to explorations by science and engineering, in that it 'makes plausible models of beings, places, and times nobody has yet encountered'.[64] In the short stories discussed earlier, electrical reanimation of the dead resembled subsequent defibrillation technologies. In *Another World*, 'concentrated light' aids microscopy, providing an interesting example of where fiction might have preceded science, by prefiguring later developments in 1893 with August Köhler's innovative use of light techniques known as 'Köhler illumination'.[65] In *Another World*, electricity is stored in a large central building called 'The Electrical Store-house', a portrayal probably inspired by the construction of the Edison Electric Light Station, which generated hydro-electric power through the Holborn Valley Viaduct in the centre of London.[66] Although the plant did not beginning operating until 1882, the visibly prominent, six-year construction project was completed in 1869, a few years before *Another World* was published (see Figure 5.2), and was a major technological development in the public mind.

Lumley's account of electricities 'secured in non-conducting pouches' and separated into 'sympathetic' and 'antagonistic' electricities, which might explode otherwise, seems informed by assorted concepts of magnetism and electrical conductivity, attraction and repulsion.[67] In 1836, John Daniell had invented the 'Daniell cell', which produced a reliable constant current, while liquid electrodes to produce electricity were devised by William Grove in 1839 and Robert Bunsen in 1842.[68] These were followed by continental innovations, including Gaston Planté's lead acid storage battery in 1859 and, just before the publication

---

63    James Clerk Maxwell, quoted in Lewis Campbell and William Garnett, *The Life of James Clerk Maxwell, with a Selection from his Correspondence and Occasional Writings and a Sketch of his Contributions to Science* (London: Macmillan and Co., 1884), 190; Janet Browne, *Charles Darwin: The Power of Place* (Princeton: Princeton University Press, 2003), 70.

64    David N. Samuelson, 'Modes of Extrapolation: The Formulas of Hard SF', *Science Fiction Studies*, 20:2 (July 1993), 191–232.

65    Lumley (1873), 63. See J. James, *Light Microscopic Techniques in Biology and Medicine* (The Hague: Martinus Nijhoff, 1976).

66    Lumley (1873), 57.

67    Lumley (1873), 57.

68    John Frederic Daniell (1790–1845); Sir William Robert Grove (1811–1896); Robert Wilhelm Eberhard Bunsen (1811–1899).

Figure 5.2 Royal procession under the Holborn Valley Viaduct (1869)[69]

of Lumley's novel, Georges Leclanché's invention of the single-fluid electric-generating battery using zinc in 1868.[70] Although usable electrical batteries would not become commercially available until the 1880s, well after the publication of *Another World*, electrical storage technologies existed sufficiently to be widely known in science and in fiction by 1873.[71] The novel's mixture of obsolete ideas about electricity with the century's newest science seems consciously designed to confuse readers and, in some cases, it succeeded. Even Bleiler concedes that 'the level of science is very high' in the novel, and that 'this is basically due to the understanding of electricity'.[72] However, as Lumley had suggested previously, the purpose of the novel was not to report his own knowledge but to play with the possibilities electricity might present.

[69]   Royal Procession under the Holborn Valley Viaduct, *Illustrated London News* (13 November 1869) © Illustrated London News Ltd/Mary Evans [www.maryevans.com/search.php; accessed 2 August 2011].

[70]   Gaston Planté (1834–1889); Georges Leclanché (1839–1882).

[71]   The first (carbon-zinc) batteries became commercially available in 1886 from the National Carbon Company, founded by the Brush Electric Company in Cleveland, Ohio; see 'Thermo-Electric Batteries', *Reader*, 6:131 (1 July 1865), 16.

[72]   Bleiler (1990), 360.

## Electrical Medicines and Treatments

Lumley presents electricity as the future of anaesthesia and neurosurgery, participating in the speculation and debates of emerging medical science. In the 1860s, safe pain relief was not yet a matter of fact but it was, as Stephanie Snow suggests, 'a touchstone for humanitarianism'.[73] *Another World* does not aim to simplify the scientific realities of pain relief and anaesthesia, nor does it work to interpret existing theories; it operates on an altogether different level. Lumley's fiction accords with the goals of 1870s research on effective pain relief and anaesthesia, a paradigm shift that historian of anaesthesia David Zuck describes as the '*acceptance of the idea* that one might safely and reversibly deliberately produce unconsciousness, hitherto an ominous sign of life-threatening illness' (author's emphasis).[74] Like the imaginary concepts used by Faraday or Maxwell, the fictions of Lumley's novel represent a form of problem-solving; Lumley's vision of anaesthesia, for example, accords with Brian Boyd's notion of fiction as a form of 'cognitive play', rather than as an earnest model of anaesthesia.[75] The novel presented, albeit in fictional form, the possibility of pain relief as an easily available, safe and manageable option; in doing so, it contributed to a particular shift in medical practices.

In the novel, pain-lulling electricity is extracted 'from a small pet-bird of pink and green plumage'.[76] The concept of electrical anaesthesia through such a juxtaposition echoes Newton's suggestion that 'bodies act upon another by the attractions of gravity, magnetism, and electricity'.[77] However, the fictitious pain-lulling electricity is 'attracted to the nerves of sensation, and the sense of feeling remains suspended during several hours while the nerves and muscles continue to function normally', an explanation that shares nineteenth-century scientific aspirations during an especially unstable period in the development of pain relief and anaesthesia.[78] The novel was published at the mid-point of nineteenth-century disputes about chemical pain relief and contemporary reviews compared the 'pain-luller' to chloroform.[79] Although the *British Medical Journal* campaigned for the use of ether in surgery, which was considered safer for patients, Queen Victoria's

---

[73]    Stephanie Snow, *Operations without Pain: The Practice and Science of Anaesthesia in Victorian Britain* (Basingstoke: Palgrave Macmillan, 2005), xiii.

[74]    David Zuck, Review: 'Operations without Pain: The Practice and Science of Anaesthesia in Victorian Britain', *Reviews in History* [www.history.ac.uk/reviews/review/573; accessed 3 December 2010].

[75]    Brian Boyd, *On the Origin of Stories: Evolution, Cognition, and Fiction* (Cambridge, MA: Belknap, 2009).

[76]    Lumley (1873), 60.

[77]    Newton, *Opticks* (1706), quoted in Patricia Fara, *Sympathetic Attractions: Magnetic Practices, Beliefs, and Symbolism in Eighteenth-Century England* (Princeton: Princeton University Press, 1996), 179–80.

[78]    Lumley (1873), 60–61; Snow (2005), xii.

[79]    *Morning Post* (1873), 211.

use of chloroform during the birth of Prince Leopold in 1853 persuaded medical authorities to endorse it as the favoured anaesthetic in 1861.[80]

Electrical pain relief on Montalluyah was critical to the other medical advances Lumley presented in the novel, for it allowed the brain to be viewed and its electrical basis to be revealed. Surgery takes place in which the patient's skull is made transparent 'by the aid of concentrated light and of an instrument called an "electric viewer", [and] the currents of electricity in the brain' are made visible.[81] Synaptic currents are revealed as 'myriads of electrical lines' and they are 'literally composed of electricity'; like 'a mathematical line', they have 'length without breadth'. The narrator specifies carefully how the brain's 'lines' (and we might think here of Faraday's 'lines of force') are not just metaphorically but actually made up of electricity. The novel attempts again to predict developments yet to happen by means of electricity, rather than just representing the goals of actual contemporary medicine. *Another World* pre-dates the 1895 discovery of X-rays by more than 20 years, but Lumley recognised the significance of electricity for developing the technology in brain surgery and neuroscience.[82]

The 1870s was a critical period in the relationship between electrical science and experimental neurology. Fiction writings about electricity, like Lumley's, contributed to the investigative environment of the period for, as Godfrey-Smith suggests, 'we engage in games of make-believe about an actual object, with the purpose of learning more about that very object'.[83] Reliable and controllable electrical currents were essential for carrying out surgical procedures in neurology. By 1873, David Ferrier (1843–1928) had already begun performing electrical stimulations of the cortex, the results of which he reported at the British Association.[84] His experiments in the 1880s at the West Riding Lunatic Asylum in Yorkshire and several London hospitals were widely publicised, alongside those of John Hughlings Jackson (1835–1911) and James Crichton-Browne (1840–1937). In *Another World*, the narrator depicts neurosurgery but he also describes the brain's physical composition and basis as electrical, with 'filaments' that are 'set in motion by the impulsion of thought'. Although the term 'filament' was used in relation to nerves by psychologist Alexander Bain in 1855, it was not associated with electricity until 1881.[85] Lumley's narrator describes the appearance of the brain's electrical 'filaments' as 'straight, spiral, and otherwise curved, and of varied

---

[80]   'The Week', *British Medical Journal*, 2 (14 December 1861), 639; see also, The Report of Chloroform Committee of Royal Medical and Chirurgical Society (1864).

[81]   Lumley (1873), 66.

[82]   X-rays were discovered accidentally by the German physicist, Wilhelm Conrad Röntgen; see Bettyann Kevles, *Naked to the Bone: Medical Imaging in the Twentieth Century* (New Brunswick: Rutgers University Press, 1997), 18.

[83]   Godfrey-Smith (2009), 106.

[84]   David Ferrier, 'Dr Ferrier's Experiments', *Athenaeum*, 2397 (4 October 1873), 440.

[85]   Alexander Bain, *Senses and Intuition*, I. ii. (1855), 14; Sylvanus P. Thompson, *Elem. Less. Electr.* (1881), 374 (OED).

length and colours', like the light bulbs that had yet to be fashioned and with which the term would become tightly linked. In the novel, the patient undergoing the operation is able to talk with the surgeons while they watch the lines moving in his brain, in response to electrical stimuli and the different subjects they discuss with him. Lumley's fictional scenario is extraordinary because it not only envisages improvements in nineteenth-century medical science but also foresees surgical procedures such as the 'awake craniotomy', which have only become possible in the last 50 years.[86] Lumley's fiction offered no specific advice on achieving the scientific actuality but, in tandem with the century's other ongoing developments, it would have expanded the *imagined* possibilities of neurosurgery.

Nineteenth-century neurosurgery was often psychosurgery, surgical operations performed on the brain to treat mental illness. In *Another World*, the procedures are portrayed as 'valuable' investigations, rather than treatments for mental illness. They are also described as rare because of the effects on patients, who could be 'insensible for some time afterwards, and felt the effects for years'. We are also told that patients were not allowed to marry afterwards, due to unspecified but 'serious consequences'.[87] Mental illness or 'madness' is mentioned only briefly in the novel and not in relation to medical treatments or innovations.[88] Like the short fictions about electricity discussed earlier, the character described as mad in *Another World* is also portrayed as a misunderstood genius; it is his untimely recognition of electricity's potential power that leads the rest of society to condemn him as a 'madman'.

The assessment of Lumley's *Another World* and his earlier novel *Sirenia* (1862) merely as 'experiments in what would later be called science fiction' underplays the literary sensation that *Another World* caused in its day.[89] Nevertheless, the novel provided many of the core elements for more celebrated late nineteenth-century fictions that portray electricity as a combination of realistic science, the supernatural and the imagined, such as Marie Corelli's *A Romance of Two Worlds* (1886), and the time travels of W.H. Hudson's *A Crystal Age* (1887) or H.G. Wells's *The Time Machine* (1895).

The precise contribution that fiction writings made to scientific developments remains difficult to gauge. However, as Kuhn recognises, 'perhaps science does not develop by the accumulation of individual discoveries and inventions' but, rather, by the influences of their wider contexts.[90] The description in popular

---

[86]    See Ketan R. Bulsara, Joel Johnson and Alan T. Villavicencio, 'Improvements in Brain Tumor Surgery: The Modern History of Awake Craniotomies', *Neurosurgery Focus*, 18:4 (April 2005), and M. Westphal, *Local Therapies for Glioma: Present Status and Future Developments* (Wein: Springer, 2003).

[87]    Lumley (1873), 67.

[88]    Lumley (1873), 73.

[89]    Middleton (2007).

[90]    Thomas Kuhn, *Scientific Revolutions* (Chicago: University of Chicago Press, 1996), 2.

novels of the contexts, experiences and potential ramifications of electricity contributed substantially to understandings and, indeed, misunderstandings of the phenomenon. At the same time, portrayals of electrical experimentation and how they were received in the press shaped wider perceptions of science, its practice and practitioners. While direct correlations between fictions about electricity and the shape of nineteenth-century sciences cannot always be established, it is clear that a lively exchange existed between them.

Several implications can be drawn from the comparison of references to electricity in canonical works and the other contemporary novels. The canonical works discussed earlier engage with ideas of electricity, but they do so by means of metaphor, which tend to rely on commonly experienced features of the phenomenon, such as volatility, speed or invisibility. The novels mentioned have a literary credibility and worth that they have retained; however, in terms of electricity, they offer engagements that are invariably oblique. The novels offer little in the way of reflection about the phenomenon, in terms of its technical properties, its contemporary significance, or its role within the broader narrative of each novel. In contrast, popular novels such as *Auriol, A Strange Story, The Coming Race* and *Another World* offered sustained and detailed depictions of electricity's properties, possibilities and repercussions. While their literary merit is sometimes debatable, their popularity makes them significant as contributions to contemporary awareness of electricity. Novels such as *Another World* prompted fascination, questioning and understandings (or misunderstandings) of scientific developments, as the reviews indicate. They were equally important elements of nineteenth-century literature, which often had more substantial influence on contemporary society than canonical works. For that reason, it is important to look beyond the canon and consider ostensibly 'marginal' novels, as well as scientists' writings, popular books and periodical writings, as part of the evaluation and reassessment of literatures in the nineteenth century. It is only through understanding more fully the pluralistic nature of nineteenth-century authorships, contexts and reading that we can achieve a more informed and balanced view of nineteenth-century literatures and culture.

# Chapter 6
# Conclusion

The primary appeal of electricity in the nineteenth century might seem, at first, to have been the pursuit of modernity; however, never far behind lay a bevy of associations with the past and contemporary readers' yearnings for those. Electrical power and its attendant technologies promised a revolutionised future, an affordable one, liberated by universal ideals of ease and convenience that could protect against the conflicts and challenges of the future, providing the foundation for the 'brilliant electric armour of the modern world'.[1] Electricity was indeed a natural phenomenon, but not one born of Nature's pastoral ease, its fruitfulness, harmony or regularity. In fact, quite the opposite. It emanated from darker and more dangerous regions, as an irrepressible and explosive discharge, an unleashed and uncontrollable violence of destructiveness and unpredictability. As Wolfgang Iser contends, 'fiction proves to be a matrix for all kinds of processes'.[2] Yet, both the fiction and the non-fiction frequently lapse into nostalgia for and misgivings about engagements with electricity that were excessively close, even Faustian, and warn of electricity's fundamentally disjointed relationship with the natural order.

Electrical technologies threatened, meanwhile, to sweep in radical changes that would disrupt traditional and long-established cadences. Daytime hours might be extended by electric light, in ways that were useful to individuals as well as commerce; but essentially those additional hours were artificial too. Like the harshly over-bright arc lights, manufactured electricity was an imitation of the normal and the real. The harnessing of electricity's power was an achievement but a potentially jarring one, a questionable advance that fundamentally distanced humankind from the timings and rhythms of the natural world, even as it supplemented man's 'natural' capacities with artificially enhanced ones. The presentation of electricity as a preternatural force confirmed and prompted further feelings of Luddite and Romantic nostalgia in the period.[3] Electricity's origins were profoundly natural but, in its modern form as electrical power, it seemed to

---

[1]  Wyndham Lewis, quoted in Michael H. Whitworth, *Modernism* (Malden: Blackwell, 2007), 16.

[2]  Wolfgang Iser, *The Fictive and the Imaginary: Charting Literary Anthropology* (Baltimore: Johns Hopkins University Press, 1993), 144.

[3]  Nichols Fox, *Against the Machine: The Hidden Luddite Tradition in Literature, Art and Individual Lives* (Washington, DC: Island Press, 2002); Martin Willis, *Mesmerists, Monsters, and Machines: Science Fiction and the Cultures of Science in the Nineteenth Century* (Kent, OH: Kent State University Press, 2006), 71.

stand in opposition to the historically natural processes of which man had always been a part.

The conceptual and physical instability of electricity as a phenomenon makes it powerfully transitive, suggestive and reflective, in ways that allow nineteenth-century writers to convey an extraordinary and even unique breadth of concerns about the past, present and future. Ostensibly naïve fictional responses that describe characters such as 'Mr Hippsley' or Elias Johns in 'Reminiscences' mask a complex network of affiliated anxieties about the psychological and physical effects of electricity and man's engagement with it. Disquiet about the pace of nineteenth-century scientific and technological change prompted portrayals of electrical experimentation as a poisonous and dehumanising process that results in the repulsive productions of the 'Tree of Knowledge', the dehumanised and intimidating *Vril-ya*, or the other-worldly artificiality of Montalluyah. The authors portray their individual experiences of electrical experimentation not simply as processes of producing a collective and useful bank of technical knowledge but, rather, as engagements with science. In doing so, they reveal both the desires and the fears that emanated from the period's fascination with electricity.

The comparison of writings in this volume reveals that electricity was repeatedly present and often relatively central in the prevailing nexus of integrated nineteenth-century literary, scientific and cultural interests, and that it was regularly implicated on a variety of levels, not just the investigative. 'Before doing physics', it has been suggested, 'one must study the essence of the physical fact'.[4] In the nineteenth century, however, conceptual encounters with electricity went considerably beyond straightforward comprehension of its physical facts. The allure of electricity pre-existed and outlasted any factual basis electricity might have had in physics and allowed considerations of the phenomenon to embrace relations with less factually distinct types of knowledge in phenomenology, philosophy and psychology, as well as the wider humanities. Understanding electricity's properties meant addressing not just provable facts but also the nature of cognition and representation, as integral features in encounters with and comprehensions of the phenomenon. The interdisciplinary nature of both electricity and responses to it were not just incidental; they considerably enabled scientists' questions about how to represent phenomena, popularisers' challenges about who could represent scientific ideas, and fiction authors' conjectures about the further underlying meanings of how we experience and relate the world. Juxtaposing the writings of all three types of sources reveals the full range of perspectives, overlaps and divergences, as well as the crucial ambiguities that prevailed within each.

Ideas about nineteenth-century electricity were formed and conveyed by being communicated; yet those processes also revealed implicit instabilities in the boundaries of science, as well as their relationships with such other spheres as literature and culture. Rather than conforming to a simple process of knowledge

---

[4]    Jean-François Lyotard, *Phenomenology*, tr. Brian Beakley (Albany: SUNY Press, 2001), 41.

being produced and consumed, wider awareness of electricity was shaped by the ways in which it was presented, by whom and where. It was by means of these decidedly multi-dimensional mechanisms that knowledge was created and communicated, and the original natural phenomenon was effectively transformed. While many readers were interested in how electricity worked, almost all of them wanted to know what it could actually do. Simultaneously, reading about electricity engaged readers emotionally and on an imaginative level. While we might expect this with poems, short stories and novels, it was also the case in readers envisaging how they might enact experiments described by Anthony Peck, in participating and seeing themselves as part of the scientific communities portrayed by Michael Faraday and William Sturgeon, and in fantasising about and seeking out the future benefits of electrical power anticipated by Robert Hunt. In these reading experiences, a blurring took place of genres, forms and disciplines, between non-fiction and fiction, between factual and imaginative, and even between scientific and literary.

Assumptions of narrative form, epistemology and audience were repeatedly questioned by nineteenth-century responses to electricity and experimentation, resulting in a veritable melange of impressions. Scientific and practical ideas existed alongside fictions about electricity. By examining them together, we gain a more complete and accurate understanding of how electricity was perceived by the period's scientists, writers and readers, in all its rich commotion. New ideas blended with old, scientific with literary, and conceptual with manifest; in studying the first, frequently, we find the second. What emerges is a variety of new ways to understand how knowledge is formed, how it develops, and the role it plays in both specialist and non-specialist explorations of complex ideas. The blend of excitement and apprehension that characterised nineteenth-century responses to electrical technologies is echoed today, in our evolving relationship with energy sources and technology. Electricity's potential to bring about revolutionary, even utopian, futures is still a dominant feature of our relationship with it today.[5] Our responses to electricity and our interest in new approaches to conserving and developing energy technologies continue to make the stuff of fantasy increasingly achievable, just as they did in the nineteenth century. Understanding responses to electricity and its accompanying technologies is not, therefore, relevant only to history, science or literature; it is also crucial to understanding how these same responses may determine our futures.

---

[5]    Jeremy Rifkin, *The Third Industrial Revolution: How Lateral Power Is Transforming Energy, the Economy, and the World* (New York: Palgrave Macmillan, 2011).

# Bibliography
## Primary

Anon., 'A Woman-Hater', *Blackwood's Edinburgh Magazine*, 121:738 (April 1877), 410–27.

Ainsworth, William Harrison, *Auriol, or the Elixir of Life* (1850; London: George Routledge and Sons, 1890).

Aldini, Giovanni, *Essai Theorique et Experimental sur le Galvanisme* (1804) [https:/.../NatureandArtifice/week6f.html; Prof. H. Beukers, Leiden University www.diagnosticarea.com/.../FES_Review.html; accessed 14 April 2010].

Anon., 'Alchemy, by An Alchemist', *Fraser's Magazine for Town and Country* 19:112 (April 1839), 446–55.

Anon., 'Another World', *Saturday Review of Politics, Literature, Science and Art*, 35:900 (25 January 1873), 123.

Anon., 'Another World' (from *Lloyd's Weekly Newspaper*), *Musical World*, 51:11 (15 March 1873), 159.

Anon., 'Another World' (from the *Morning Post*), *Musical World*, 51:14 (5 April 1873), 211.

Anon., 'Another World', *Musical World*, 51:14 (5 April 1873), 220.

Anon., 'Another World' (from *The Spiritualist*, 15 March), *Musical World*, 51:17 (26 April 1873), 263.

Anon., 'Another World' (from *The Echo*), *Musical World*, 51:18 (3 May 1873), 279.

Anon., 'Another World', *Macmillan's Magazine*, 28:164 (June 1873), 140.

Anon., Review: 'Auriol, and other Tales', *Critic*, 9:233 (15 December 1850), 591.

Babbage, Charles, *Reflections on the Decline of Science in England* (London: B. Fellowes and J. Booth, 1830).

Bachhoffner, G. H., *A Popular Treatise on Voltaic Electricity and Electro-Magnetism* (London: Simpkin and Marshall, 1838).

Barrett, W.F., Edmund Gurney and Frederic W.H. Myers, 'Thought Reading', *Nineteenth Century*, 11:64 (June 1882), 890–901.

Beard, G. and A. Rockwell, *On the Medical and Surgical Use of Electricity* (New York: William Wood and Company, 1891).

Beccaria, Giambatista, *A Treatise upon Artificial Electricity* (London, 1780).

Black, Adam and Charles (Firm), *An Attempt to Simplify the Theories of Electricity and Light* (Edinburgh: Adam and Charles Black, 1834).

Boyle, Robert, Hon., *Some Motives and Incentives to the Love of God, in a Letter to a Friend*, 4th edn ([1659] London, 1665).

Brewster, David, *The Life of Sir Isaac Newton* (London: John Murray, 1831).

———— *Letters on Natural Magic: Addressed to Sir Walter Scott* (London: John Murray, 1832).

Brewster, David, Richard Taylor and Richard Phillips, eds, *London and Edinburgh Philosophical Magazine and Journal of Science*, vol. 6 (London: Taylor & Francis, 1835).

Bridgeman, John V., 'Shocks', *Train: A First-Class Magazine*, 4:20 (August 1857), 111–15.

Brontë, Anne, *The Tenant of Wildfell Hall* (London: T.C. Newby, 1848).

Brontë, Charlotte, *Shirley* (1849; repr. London: John Murray, 1929).

———— *Villette* (London: Smith, Elder and Co., 1853).

Brontë, Emily, *Wuthering Heights* (London: Thomas Cautley Newby, 1847).

Bulwer-Lytton, Edward, Baron, 'A Strange Story', *All the Year Round*, 6:128 (October 1861), 25–9 to 6:150 (March 1862), 553–8.

———— *A Strange Story* (Boston: Gardner A. Fuller, 1862).

———— 'On Certain Principles in Art in Works of Imagination', *Miscellaneous Prose Works*, vol. 3 (London: R. Bentley, 1868).

————*The Coming Race*, ed. Matthew Sweet (London: Hesperus Press, 2007).

———— *The Coming Race*, ed. Peter W. Sinnema (1871; repr. Peterborough: Broadview Press, 2008).

Campbell, Lewis and William Garnett, *The Life of James Clerk Maxwell, with a Selection from his Correspondence and Occasional Writings and a Sketch of his Contributions to Science* (London: Macmillan and Co., 1884).

Carlyle, Thomas, *Sartor Resartus: The Life and Opinions of Herr Teufelsdröckh in Three Books*, ed. Rodger L. Tarr and Mark Engel (1831; repr. Berkeley: University of California Press, 2000).

———— 'Signs of the Times' [1829], in Himmelfarb (2007), 31–49.

Chambers, William and Robert Chambers, *Chambers's Information for the People*, vol. 2 (Edinburgh: W. and R. Chambers, 1842).

Clark, P., 'The Symmes Theory of the Earth', *Atlantic Monthly*, 31 (April 1873), 471–80.

Collins, Wilkie, *The Moonstone* (New York: Harper and Brothers, 1868).

———— *Armadale*, vol. 2 (London: Smith, Elder and Co., 1869).

———— *Man and Wife* (Leipzig: Bernhard Tauchnitz, 1870).

————*No Name*, ed. Mark Ford (1862; repr. London: Penguin, 1994).

Anon., 'Columbia', *Saturday Review of Politics, Literature, Science and Art*, 35:905 (1 March 1873), 289.

Crosse, Andrew, 'Electrical Society', *Literary Gazette*, 1097 (27 January 1838), 54–6.

———— 'Electrical Society', *Annals of Electricity, Magnetism and Chemistry*, vol. 2 (January–June 1838).

Dalton, John, *New System of Chemical Philosophy*, vol. 1 (Manchester: Printed by S. Russell, 125, Deansgate, for R. Bickerstaff, Strand, London, 1808–1827).

Daniell, Frederic, 'Professor Daniel's Additional Observations', *Annals of Electricity Magnetism and Chemistry*, 1:18 (April 1836), 102–3.

————— 'Sixth Letter on Voltaic Combinations', *Annals of Electricity, Magnetism and Chemistry* 10:57 (March 1843), 232–40 and 10:58 (April 1843), 241–53.

Darcy, Henry, *Les Fontaines Publiques de la Ville de Dijon*, tr. 'The Public Fountains of the Town of Dijon' (Paris: Dalmont, 1856).

Darwin, Erasmus, *Temple of Nature, Or, The Origin of Society: A Poem, with Philosophical Notes*, ed. Martin Priestman [www.rc.umd.edu/editions/darwin_temple/; accessed 20 July 2009].

Demonferrand, Jean Firmin, tr. James Cumming, *A Manual of Electrodynamics* (Cambridge: J.J. Deighton, 1827).

Dickens, Charles, *The Old Curiosity Shop: A Tale* (London: Chapman and Hall, 1841).

————— *The Life and Adventures of Martin Chuzzlewit*, vol. 1 (London: Chapman and Hall, 1844).

————— *Dombey and Son* (London: Bradbury and Evans, 1848).

————— 'A Physician's Ghosts', *All the Year Round*, 15 (August 1859).

Anon., 'Doctor Beroni's Secret (I)', *Temple Bar*, 72 (October 1884), 187–208.

Anon., 'Doctor Beroni's Secret (II)', *Temple Bar*, 72 (November 1884), 339–60.

Dowling, Richard, 'An Anachronism from the Tomb', *Tinsley's Magazine*, 24 (February 1879), 147–66.

Anon., 'Dr. Duncan on "Catalepsy"', *The Lancet*, 2 (1830), 277–84.

Eberle, John, *Treatise on the Practice of Medicine*, vol. 2 (Philadelphia: Grigg and Elliot, 1835).

Anon., 'Electrical Society', *Literary Gazette*, 1092 (23 December 1837), 818.

Anon., 'Electricity', *Saturday Magazine*, 6:169 (February 1835), 68–9.

Anon., 'Electricity', *Saturday Magazine*, 13:399 (September 1838), 111–12.

Anon., 'Electricity', *Chambers's Journal of Popular Literature, Science and Arts*, 123 (10 May 1856), 303–4.

Anon., Review: 'Electricity; its Nature, Operation, and Importance in the Phenomena of the Universe' by William Leithead, *British Magazine*, 12 (December 1837), 675.

Anon., 'Electricity of Animal Life', *The Penny Magazine of the Society for the Diffusion of Useful Knowledge*, 8:439 (2 February 1839), 48.

Anon., 'Electricity and Vegetation', *World of Science*, 1:10 (14 December 1867), 140.

Eliot, George, *Scenes of Clerical Life*, vol. 2 (Edinburgh and London: William Blackwood and Sons, 1858).

————— *Adam Bede*, vol. 2 (Leipzig: Bernhard Tauchnitz, 1859).

————— *Daniel Deronda*, vols 3 and 4 (London: William Blackwood and Sons, 1876).

————— *Middlemarch*, ed. Rosemary Ashton (London: Penguin, 2003).

Faraday, Michael 'Experimental Researches in Electricity', *Philosophical Transactions of the Royal Society of London*, 122 (1832), 125–62.

——— 'Additional Observations Respecting the Magneto-Electric Spark and Shock', *Philosophical Magazine and Journal of Science* (London and Edinburgh, December 1834).

——— *Experimental Researches in Electricity*, vol. 1 (London: Richard John and Edward Taylor, 1839).

——— *Experimental Researches in Electricity*, series 1–14 [*Phil. Trans.*, 1831–1838] (London: Bernard Quaritch, 1839).

——— *On the Liquefaction and Solidification of Bodies Generally Existing as Gases* (London: R. and J. E. Taylor, 1845).

——— 'Experimental Researches in Electricity' [1852], in Otis (2002), 55–9.

——— 'Observations on Mental Education' [1854], in Jenkins (2008), 200–12.

Ferguson, Robert M., *Electricity* (Edinburgh: William and Robert Chambers, 1867).

Ferrier, David, 'Dr Ferrier's Experiments,' *Athenaeum*, 2397 (4 October 1873), 440.

Fortune, Robert, *Two Visits to the Tea Countries of China and the British Tea Plantations in the Himalaya: with a Narrative of Adventures, and a Full Description of the Culture of the Tea Plant, the Agriculture, Horticulture, and Botany of China*, vols 1 and 2 (London: John Murray, 1853).

Fox, Robert Were, 'Experiments on the Influence of Electrical Action upon Clay,' *Annals of Electricity, Magnetism and Chemistry* 2:7 (January 1838), 54.

Frick, Joseph, *Physical Technics: or, Practical Instructions for Making Experiments in Physics and the Construction of Physical Apparatus with the Most Limited Means* (Philadelphia: J.B. Lippincott and Co., 1861).

Froude, James, *The Nemesis of Faith* (London: J. Chapman, 1849).

Gale, T.G., *Electricity, or Ethereal Fire* (Troy: Moffit and Lyon, 1802).

Anon., 'General Intelligence', *John Bull*, 1 (8 July 1848), 439.

Gladstone, John Hall, 'Faraday: Review of The Life and Letters of Faraday by Dr. Bence Jones', *Nature* (17 February 1870), 403.

Halley, Edmond, 'An Account of the Cause of the Change of the Variation of the Magnetic Needle; with an Hypotheses of the Structure of the Internal Parts of the Earth,' *Philosophical Transactions of Royal Society of London*, 16 (1692), 563–78.

Harris, William Snow, Sir, *Rudimentary Electricity, being a Concise Exposition of the General Principles of Electrical Science* (London: John Weale, 1848).

——— *Rudimentary Magnetism* (London, 1850).

——— *Rudimentary Treatise on Galvanism, and the General Principles of Animal and Voltaic Electricity* (London, 1856).

——— *On a General Law of Electrical Discharge* [1856?]; *A Treatise on Frictional Electricity* (London, 1867).

Hawthorne, Nathaniel, 'A Birth-Mark', *The Pioneer* (March 1843), repr. in *Mosses from an Old Manse* (London, 1846).

'Hermes', 'Summary of Science', *Critic*, 10:257 (15 December 1851), 609.

——— 'Science and Inventions' (series), *Critic*, 11:258 (1 January 1852), 46–7; 11:261 (16 February 1852), 102–3; 14:336 (2 April 1855), 169–70; 14:347 (15 September 1855), 457–8.

———— 'Scientific Summary' (series), *Critic*, 14:340 (1 June 1855), 264–5; 14:341 (15 June 1855), 288–9; 14:348 (1 October 1855), 488.

———— 'Planetary Life', *All the Year Round*, 11:257 (1 November 1873), 5–9; 11:260 (22 November 1873), 84–8; 12:283 (2 May 1874), 53–5.

———— 'Planetary Life', *Musical World*, 51:46 (15 November 1873), 770–71; 51:47 (22 November 1873), 783; 52:19 (9 May 1874), 302.

Hinton, James, 'Physiological Riddles. I.—How We Act', *Cornhill Magazine*, 2:1 (July 1860), 21–32; 'II.—Why We Grow', *Cornhill Magazine*, 2:2 (August 1860), 167–74; 'III.—Living Forms', *Cornhill Magazine*, 2:3 (September 1860), 313–25; 'IV.—Conclusion', *Cornhill Magazine*, 2:4 (October 1860), 421–31.

Horne, Richard, *The Poor Artist; or, Seven Eye-Sights and One Object* (London: John Van Voorst, 1850).

Hunt, Robert, *Researches on Light: An Examination of all the Phenomena Connected with the Chemical and Molecular Changes* (London: Longman, Brown, Green, and Longmans, 1844).

———— *The Poetry of Science, or Studies of the Physical Phenomena of Nature* (London: Reeve, Benham and Reeve, 1848).

Hutton, Charles, *A Philosophical and Mathematical Dictionary* (London: Printed for the Author, 1815).

Jackson, J.H., 'The Factors of Insanities', *Medical Press Circular*, 108 (1894), 615–19.

Johns, Richard, 'Mr. Hippsley, The Electrical Gentleman', *Bentley's Miscellany*, 4 (July 1838), 374–82.

Joyce, Jeremiah, *Scientific Dialogues: Intended for the Instruction and Entertainment of Young People; in which the First Principles of Natural and Experimental Philosophy are Fully Explained* (1805; repr. London: William Tegg, 1842). [Reprinted by William Tegg in 1868 as *Scientific Dialogues: Intended as an Easy Introduction to Natural and Experimental Philosophy for Young Persons and All Who by Self-Instruction Desire to Understand the Principles of Scientific Truth.*]

Kahlbaum, Georg W.A. and Francis Vernon Darbishire, eds, *The Letters of Faraday and Schoenbein, 1836–1862* (London: Williams and Norgate, 1899).

Kane, Robert, 'Contributions to the Chemical History of Palladium, Communicated by Francis Bailey', *Annals of Electricity, Magnetism and Chemistry*, 10:58 (April 1843), 253–71.

La Beaume, Michael, *Remarks on the History and Philosophy but Particularly on the Medical Efficacy of Electricity in the Cure of Nervous and Chronic Disorders* (London: F. Warr, 1820).

Leithead, William, *Electricity: Its Nature, Operation and Importance in the Phenomena of the Universe* (London: Longman, Orme, Brown, Green, and Longmans, 1837).

Lewes, George Henry, *The Physiology of Common Life*, vol. 2 (New York: Appleton, 1867).

Anon., 'Liquefaction of Oxygen', *The Times*, 29134 (25 December 1877), 4.

Lumley, Benjamin, *Another World, or, Fragments from the Star City of Montalluyah, by 'Hermes'* (London, 1873).

Matthews, Brander, ed. *The Oxford Book of American Essays* (New York: Oxford University Press, 1914).

Maxwell, James Clerk, *Experiments on Colour, as Perceived by the Eye, with Remarks on Colour Blindness* (Edinburgh: Neill and Company, 1855).

────── *Theory of Heat* (London, 1871).

────── 'On Colour Vision', *Royal Institution Proceedings*, 6 (1872), 260–71.

────── 'Are there Real Analogies in Nature?' (February 1856), in Campbell and Garnett (1884), 235–44.

────── 'Lectures to Women on Physical Science (I)', in Campbell and Garnett (1884).

────── 'A Problem of Dynamics (February 19th, 1854)', in Campbell and Garnett (1884).

────── 'Reflex Musings: Reflection from Various Surfaces', in Campbell and Garnett (1884).

────── 'A Tyndallic Ode', in Campbell and Garnett (1884).

────── 'Valentine by a ♂ Telegraph Clerk to a ♀ Telegraph Clerk', in Campbell and Garnett (1884).

────── 'On Faraday's Lines of Force' [1855–1856], in Niven (1965).

────── 'General Considerations concerning Scientific Apparatus', in Niven (1965), 505–22.

────── 'Introductory Lecture on Experimental Physics', Cambridge, undated, in Niven (1965).

Mill, J.S., 'The Spirit of the Age', *Examiner*, 3:2 (13 March 1831), 162–3.

Mitchell, James, ed., *Dictionary of the Mathematical and Physical Sciences* (London: Printed for Sir Richard Phillips, 1823).

Anon., 'More about Electricity', *Chambers's Journal of Popular Literature, Science and Arts*, 790 (15 February 1879), 107–10.

Anon., 'Morse's Magnetic Telegraph', *John Bull*, 1:230 (6 July 1844), 426.

Anon., 'Mr Wilson's Entertainments', *Bell's Life in London and Sporting Chronicle* (21 May 1848), 3.

Anon., 'Music in Another World (from The Sunday Times)', *Musical World*, 51:15 (12 April 1873), 227.

Noad, Henry Minchin, *A Course of Eight Lectures on Electricity, Galvanism, Magnetism, and Electro-Magnetism* (London: Scott, Webster and Geary, 1839).

────── *Lectures on Electricity: Comprising Galvanism, Magnetism, Electro-Magnetism, Magneto- and Thermo-Electricity* (London: George Knight and Sons, 1844).

Anon., 'Obituary: Edward Turner, M.D., F.R.S.', *Gentleman's Magazine* (April 1837).

Anon., 'Odds and Ends', *John Bull*, 854 (24 April 1837), 194.

Anon., 'On the Origin of the Black Art, or Magic', *The Penny Satirist*, 375 (22 June 1844), 3.

Anon., 'Our Electric Selves', *Punch*, 26:661 (11 March 1854), 106.

Peck, Anthony, 'Papers on Popular Science: I. Electricity', *Reynolds's Miscellany*, 1:8 (26 December 1846), 125–6.

——— 'II. Electrical Machines', *Reynolds's Miscellany*, 1:9 (2 January 1847), 139–40.

——— 'XIII. Thermo-Electricity – Electricity of Steam – Electro-Vegetation', *Reynolds's Miscellany*, 1:27 (8 May 1847), 428–31.

——— 'On Mechanics. I. Matter and its Properties.—Force.—Attraction of Gravitation', *Reynolds's Miscellany*, 2:28 (15 May 1847), 14–15.

Peters, Dishley, 'Another World', *Musical World*, 51:27 (5 July 1873), 451.

Phillips, Richard, Sir (1767–1840), *An Easy Grammar of Natural and Experimental Philosophy: For the Use of Schools* (London: Printed for Richard Phillips, 1807).

Phipson, T.L., 'Electricity at Work', *Macmillan's Magazine*, 6:32 (June 1862), 163–9.

Poe, Edgar Allan, 'Twice-Told Tales: A Review', *Graham's Magazine* (April 1842); repr. in *Edgar Allan Poe: Essays and Reviews*, ed. G.R. Thompson (New York: The Library of America, 1984), 568–9.

——— 'The Philosophy of Composition' in Matthews (1914), 99–113.

Anon., 'Popular Science', *The Lady's Newspaper*, 5 (30 January 1847), 102.

Priestley, Joseph, *A Familiar Introduction to the Study of Electricity* (London: J. Johnson, 1786).

Anon., 'Reminiscences of a Medical Student', *New Monthly Magazine and Humorist*, 64:253 (January 1842), 123–41.

Anon., Reviews, *John Bull*, 3:053 (14 June 1879), 377.

Anon., Review: 'Rudimentary Electricity; being a Concise Exposition of the General Principles of Electrical Science, and the Purposes to which it has been Applied', *Edinburgh Review*, 90:182 (October 1849), 434–72.

Anon., Review: 'Rudimentary Works for Beginners', *Athenaeum*, 1196 (28 September 1850), 1025.

Reynolds, G.W., 'To Our Readers', *Reynolds's Miscellany of Romance, General Literature, Science, and Art*, 1:1 (7 November 1846), 16.

Ritchie, William, *Experimental Researches in Voltaic Electricity and Electro-Magnetism* (London: Richard Taylor, 1832).

Anon., 'Royal Colosseum, Regent's Park', *Sharpe's London Magazine of Entertainment and Instruction*, 28 (July 1858), 166.

Rutherford, W.M., 'Lectures on Experimental Physiology: Lecture IV: Innervation', *The Lancet*, 1 (1871), 563–7.

Sachs, Julius and Sydney Howard Vines, *Text-Book of Botany, Morphological and Physiological*, 2nd edn (Oxford: Clarendon Press, 1882).

Anon., 'Science', *Westminster Review*, 65:127 (January 1856), 254.

Scott, Walter, Sir, *Letters on Demonology and Witchcraft: Addressed to J.G. Lockhart* (London: Murray, 1830).

Sherman, M.L. and William F. Lyon, *The Hollow Globe: The World's Agitator and Reconciler a Treatise on the Physical Conformation of the Earth* (Chicago: Religio-Philosophical Publishing House, 1871).

Anon., 'Simple Electrical Machines', *Peter Parley's Annual: A Christmas and New Year's Present for Young People* [Date Unknown: 19th Century UK Periodicals; Gale Document Number: DX1901717113].

Anon., 'Sirenia', *Saturday Review*, 13:336 (5 April 1862), 390.

Smee, Arthur, *Instinct and Reason: Deduced from Electro-Biology* (London: Reeve and Benham, 1850).

Smiles, Samuel, *Self-Help: With Illustrations of Character and Conduct* (1859; repr. London: Routledge/Thoemmes Press, 1997).

Smith, Albert, 'The Adventures of Mr. Ledbury', *Bentley's Miscellany*, 12 (July 1842).

Anon., 'Spiritism and its Raps', *Orchestra*, 17:421 (20 October 1871), 42.

Sturgeon, William, *Recent Experimental Researches in Electro-Magnetism, and Galvanism: Comprising an Extensive Series of Curious Experiments, and their Singular and Interesting Results; showing that Electro-Magnetic Action may be Developed and Modified by Processes not Genrally* [sic] *Known.--With some Practical and Theoretical Observations on that Department of Science* (London: Sherwood, Gilbert and Piper, 1830).

———— 'Address to the General Meeting of the London Electrical Society', *Annals of Electricity, Magnetism and Chemistry*, 2:11 (October 1837), 64–72.

———— Review: 'Electricity; its Nature, Operation, and Importance in the Phenomena of the Universe by William Leithead', *Annals of Electricity, Magnetism and Chemistry*, 2:11 (October 1837), 77.

———— 'Miscellaneous Articles', *Annals of Electricity, Magnetism and Chemistry*, 2:11 (May 1838), 95.

———— 'Lamination of Clay by Electricity', *Annals of Electricity, Magnetism and Chemistry*, 3:14 (August 1838), 159–61.

———— 'Prize Volumes of the Annals of Electricity, & c.', *Annals of Electricity, Magnetism and Chemistry*, 6:31 (January 1841), 80–81.

[Sturgeon, William], 'Award of Prizes', *Annals of Electricity, Magnetism and Chemistry*, 6:36 (June 1841), 512.

Anon., 'Table Talk', *Once a Week*, 2:34 (22 August 1868), 158.

Thackeray, William Makepeace, *The History of Pendennis* (Leipzig: Bernard Tauchitz, 1849).

Anon., 'The Coming Race', *Saturday Review of Politics, Literature, Science and Art*, 31:813 (27 May 1871), 674.

Anon., 'The Coming Race', *Athenaeum*, 2274 (27 May 1871), 649.

Anon., 'The Electric Light', *World of Science*, 1:13 (4 January 1868), 181–2.

Anon., 'The Electrical Minister', *Punch*, 10 (28 March 1846), 145.

Anon., 'The Gifts of Science to Art,' *Dublin University Magazine*, 36:211 (July 1850), 1–3.

Anon., 'The Origin of Electricity', *Saturday Review of Politics, Literature, Science and Art*, 17:438 (19 March 1864), 348–50.

Anon., 'The Parisians', *Edinburgh Review*, 139:284 (April 1874), 383.

Anon., Review: 'The Physiology of Common Life', *The Crayon*, 6:3 (March 1859), 71–3.

Anon., Review: 'The Poetry of Science; or, Studies of the Physical Phenomena of Nature', *North British Review*, 13:25 (May 1850), 117–58.

Anon., 'The Tree of Knowledge', *Dublin University Magazine*, 41:246 (June 1853), 663–75.

Anon., 'The Universality of Electricity', *Punch*, 35:902 (30 October 1858), 694.

Anon., 'The Week', *British Medical Journal*, 2 (14 December 1861), 639.

Anon., 'Thermo-Electric Batteries', *Reader*, 6:131 (1 July 1865), 16.

Thomson, Thomas and William Blackwood, *An Outline of the Sciences of Heat and Electricity* (London: Baldwin and Cradock; William Blackwood, Edinburgh, 1830).

Turner, Edward, *Elements of Chemistry: Including the Recent Discoveries and Doctrines of the Science* (Edinburgh: William Tait and Charles Tait, 1827).

Tyndall, John, *Faraday as Discoverer* (London, 1868).

—— *Essays on the Use and Limit of the Imagination in Science* (London: Longmans, Green, and Co, 1870).

—— *Notes of a Course of Nine Lectures on Light* (London: Longmans, Green, 1870).

—— *Notes of a Course of Seven Lectures on Electrical Phenomena and Theories* (London, 1870).

—— *Fragments of Science for Unscientific People: A Series of Detached Essays, Lectures and Reviews* (New York: D. Appleton and Company, 1871).

—— *Light and Electricity* (London: D. Appleton and Company, 1871).

—— *Lessons in Electricity at the Royal Institution, 1875–6* (London: Spottiswoode and Co., 1876).

—— *Fragments of Science for Unscientific People* (1868; repr. Ann Arbor: University of Michigan Library, 2006).

—— 'Lessons in Electricity', *Christmas at the Royal Institution: An Anthology of Lectures*, World Scientific Publishing Co. Pte. Ltd

Anon., 'Upas, The Poison Tree of Java', *Encyclopaedia Britannica*, 17 (1823), 58.

Volta, Alessandro, *Of the Method of Rendering Very Sensible the Weakest Natural or Artificial Electricity* (London: J. Nichols, 1782).

Anon., 'What is Electricity?' *Once a Week*, 4:84 (2 February 1861), 163–5.

Whewell, William, *Philosophy of the Inductive Sciences*, 1 ([1840] London: Routledge/Thoemmes Press 1996).

Wilkins, Charles, Sir, *Bhagvat-Geeta, or Dialogues of Kreeshna and Arjoon* (London: Nourse, 1785).

Wilson, Benjamin, *A Short View of Electricity* (London, 1780).

Wilson, William, *A Little Earnest Book on a Great Old Subject* (London: Darton and Co., 1851).

Wordsworth, William, 'The Tables Turned', in Wordsworth and Coleridge (1798), 186–8.

Wordsworth, William and Samuel Taylor Coleridge, *Lyrical Ballads* (London: J. and A. Arch, 1798).

[www.worldscibooks.com/etextbook/6583/6583_chap04.pdf; accessed 14 June 2010].

# Bibliography
## Secondary

Archer, Mary D. and Christopher D. Haley, *The 1702 Chair of Chemistry at Cambridge: Transformation and Change* (Cambridge: Cambridge University Press, 2005).

Arnold, Matthew, *Culture and Anarchy: An Essay in Social and Political Criticism* (London: Smith, Elder and Co., 1869).

Ashall, Frank, *Remarkable Discoveries!* (Cambridge: Cambridge University Press, 1994).

Aspray, William and Philip Kitcher, eds, *History and Philosophy of Modern Mathematics* (Minneapolis: University of Minnesota Press, 1988).

Baldwin, George Benedict, *Beyond Nationalization: The Labor Problems of British Coal* (Cambridge, MA: Harvard University Press, 1955).

Barton, Ruth, 'Just before Nature: The Purposes of Science and the Purposes of Popularization in Some English Popular Journals of the 1860s', *Annals of Science* 55 (1998), 18–26.

Beasley, Edward, *The Victorian Reinvention of Race: New Racisms and the Problem of Grouping in the Human Sciences* (London: Routledge, 2010).

Beaumont, Matthew, 'The Party of Utopia: Utopian Fiction and the Politics of Readership, 1880–1900', in Griffin and Moylan (2007), 163–82.

Beer, Gillian, 'Problems of Description in the Language of Discovery', in Levine (1987), 35–58.

——— *Open Fields: Science in Cultural Encounter*, 2nd edn (Oxford: Oxford University Press, 2006).

——— *Darwin's Plots: Evolutionary Narrative in Darwin, George Eliot and Nineteenth-Century Fiction* (London: Routledge and Kegan Paul, 1983), and 3rd edn (Cambridge: Cambridge University Press, 2009).

Beetham, Margaret and Kay Boardman, *Victorian Women's Magazines: An Anthology* (Manchester: Manchester University Press, 2001).

Beik, William and Gerald Strauss, 'The Dilemma of Popular History', *Past and Present*, 141 (1993), 207–15.

Ben-Chaim, Michael, *Experimental Philosophy and the Birth of Empirical Science: Boyle, Locke and Newton* (Aldershot: Ashgate, 2004).

Bendixen, Alfred and James Nagel, *A Companion to the American Short Story* (Oxford: Wiley-Blackwell, 2010).

Berman, Russell, 'Popular Culture and Populist Culture', *Telos*, 82 (1991), 59–70.

Bernal, John Desmond, *Science and Industry in the Nineteenth Century* (London: Routledge, 2006).

Berrios, German E., 'Positive and Negative Symptoms and Jackson: A Conceptual History,' *Archives General Psychiatry*, 42:1 (January 1985), 95–7.

Bertucci, Paola and Giuliano Pancaldi, *Electric Bodies: Episodes in the History of Medical Electricity* (Bologna: Università di Bologna, 2001).

Binkley, Robert Cedric, *Realism and Nationalism, 1852–1871* (New York: Harper and Brothers, 1935).

Bleiler, Everett F., *Science-Fiction, the Early Years* (Kent, OH: Kent State University Press, 1990).

Boase, G.C., 'Noad, Henry Minchin (1815–1877)', rev. Iwan Rhys Morus, *Oxford Dictionary of National Biography* (Oxford University Press, 2004) [http://ezproxy.ouls.ox.ac.uk:2117/view/article/20214; 1 accessed June 2011].

Bonchev, Danail and D.H. Rouvray, *Chemical Topology: Introduction and Fundamentals* (Amsterdam: Gordon and Breach, 1999).

Bortoft, Henri, *The Wholeness of Nature: Goethe's Way toward a Science of Conscious Participation in Nature* (Morpeth: Lindisfarne Press, 1996).

Botting, Fred, *Making Monstrous: Frankenstein, Criticism, Theory* (Manchester: Manchester University Press, 1991).

———— 'Metaphors and Monsters', *Journal for Cultural Research*, 7:4 (2003), 339–65.

Bouchard, Donald F. and Sherry Simon, eds, *Language, Counter-Memory, Practice* (Ithaca, NY: Cornell University Press, 1977).

Bowler, Peter J., Presidential Address, 'Experts and Publishers: Writing Popular Science in Early Twentieth-Century Britain, Writing Popular History of Science Now', *British Journal for the History of Science*, 39:2 (June 2006), 159–87.

Bown, Nicola, Carolyn Burdett and Pamela Thurschwell, *The Victorian Supernatural* (Cambridge: Cambridge University Press, 2004).

Boyce, Mary, *A History of Zoroastrianism* (Leiden: Brill, 1975).

Boyd, Brian, *On the Origin of Stories: Evolution, Cognition, and Fiction* (Cambridge, MA: Belknap, 2009).

Boyer, Carl Benjamin, *A History of Mathematics*, rev. Uta C. Merzbach, 2nd edn (New York: Wiley, 1989).

Brake, Laurel and Julie F. Codell, eds, *Encounters in the Victorian Press: Editors, Authors, Readers* (Basingstoke: Palgrave Macmillan, 2005).

Brake, Laurel, Aled Jones and Lionel Madden, *Investigating Victorian Journalism* (Basingstoke: Macmillan, 1990).

Brantlinger, Patrick, 'Race and the Victorian Novel', in David (2001), 149–68.

Brewer, John and Roy Porter, eds, *Consumption and the World of Goods* (London: Routledge, 1993).

Bristow, Joseph, ed., *The Cambridge Companion to Victorian Poetry* (Cambridge: Cambridge University Press, 2000).

Brock, W.H.N., 'Turner, Edward (1796–1837)', *Oxford Dictionary of National Biography*, (Oxford University Press, September 2004; online edition, January 2009) [http://ezproxy.ouls.ox.ac.uk:2117/view/article/27848; accessed 10 July 2010].

Brock, W.H.N., D. McMillan and R.C. Mollan, eds, *John Tyndall: Essays on a Natural Philosopher* (Dublin: Royal Dublin Society, 1981).

Broks, Peter, *Understanding Popular Science* (Maidenhead: Open University Press, 2006).

Brown, Andrew, 'Lytton, Edward George Earle Lytton Bulwer, First Baron Lytton (1803–1873)', *Oxford Dictionary of National Biography* (Oxford University Press, 2004) [http://ezproxy.ouls.ox.ac.uk:2117/view/article/17314; accessed 2 November 2010].

Brown, Daniel, 'Victorian Poetry and Science', in Bristow (2000), 137–58.

Browne, Janet, *Charles Darwin: The Power of Place* (Princeton: Princeton University Press, 2003).

Browning, Logan, 'The Irregular Publication of "Regular Habits": Dr. Charles Julius Roberts and "Bentley's Miscellany"', *Victorian Periodicals Review*, 23:2, Wellesley Index Special Issue (Summer 1990), 60–64.

Bulsara, Ketan R., Joel Johnson and Alan T. Villavicencio, 'Improvements in Brain Tumor Surgery: the Modern History of Awake Craniotomies', *Neurosurgicaly Focus*, 18:4 (April 2005), 1–3.

Caleb, Amanda Mordavsky, *(Re)creating Science in Nineteenth-Century Britain* (Newcastle: Cambridge Scholars, 2007).

Cantor, Geoffrey, 'The Rhetoric of Experiment', in Gooding et al. (1989), 159–80.
––––––– *Michael Faraday, Sandemanian and Scientist: A Study of Science and Religion in the Nineteenth Century* (Basingstoke: Macmillan 1991).

Cantor, Geoffrey and M.J.S. Hodge, *Conceptions of Ether: Studies in the History of Ether Theories, 1740–1900* (Cambridge: Cambridge University Press, 1981).

Cantor, Geoffrey and Sally Shuttleworth, eds. *Science Serialized: Representation of the Sciences in Nineteenth-Century Periodicals* (Cambridge, MA: MIT Press, 2004).

Cantor, Geoffrey, David Gooding and Frank A.J.L. James, *Faraday* (Basingstoke: Macmillan, 1991).

Cantor, Geoffrey, Gowan Dawson, Graeme Gooday, Richard Noakes, Sally Shuttleworth and Jonathan R. Topham, eds. *Science in the Nineteenth-Century Periodical: Reading the Magazine of Nature* (Cambridge: Cambridge University Press, 2004).

Carcione, José M., *Wave Fields in Real Media: Wave Propagation in Anisotropic, Anelastic and Porous Media* (Oxford: Pergamon, 2001).

Chadwick, Owen, *An Ecclesiastical History of England: The Victorian Church* (Oxford: Oxford University Press, 1966).

Chan, Winnie, *The Economy of the Short Story in British Periodicals of the 1890s* (London: Routledge, 2007).

Christ, Carol T. and John O. Jordan, eds. *Victorian Literature and the Victorian Visual Imagination* (Berkeley: University of California Press, 1995).

Christie, John and Sally Shuttleworth, eds. *Nature Transfigured: Science and Literature, 1700–1900* (Manchester: Manchester University Press, 1989).

Christie, John R.R., 'The Development of the Historiography of Science', in Olby et al. (1990), 5–22.

Cogman, Peter, *Narration in Nineteenth-Century French Short Fiction: Prosper Mérimée to Marcel Schwob* (Durham: University of Durham, 2002).

Cohen-Cole, Jamie, 'The Reflexivity of Cognitive Science: the Scientist as Model of Human Nature', *History of the Human Sciences*, 18 (November 2005), 107–39.

Cooter, Roger and Stephen Pumfrey, 'Separate Spheres and Public Places: Reflections on the History of Science Popularization and Science in Popular Culture', *History of Science*, 32 (September 1994), 237–67.

Crawford, Robert, ed. *Contemporary Poetry and Contemporary Science* (Oxford: Oxford University Press, 2006).

Cronin, Richard, Alison Chapman and Antony H. Harrison, eds. *A Companion to Victorian Poetry* (Oxford: Blackwell, 2002).

Crowe, Michael J., 'Ten Misconceptions about Mathematics and Its History', in Aspray and Kitcher (1988), 260–77.

Dale, Peter Allan, *In Pursuit of a Scientific Culture: Science, Art and Society in the Victorian Age* (Madison: University of Wisconsin Press, 1989).

Dames, Nicholas, *The Physiology of the Novel: Reading, Neural Science, and the Form of Victorian Fiction* (Oxford: Oxford University Press, 2007).

Darrigol, Olivier, *Electrodynamics from Ampère to Einstein* (Oxford: Oxford University Press, 2002).

Daston, Lorraine and Peter Galison, *Objectivity* (New York: Zone Books, 2007).

David, Deirdre, ed. *The Cambridge Companion to the Victorian Novel* (Cambridge: Cambridge University Press, 2001).

———— 'Sensation and the Fantastic in the Victorian Novel', in David (2001), 192–211.

Dawson, Gowan, 'The Cornhill Magazine and Shilling Monthlies in Mid-Victorian Britain', in Cantor et al. (2004), 123–50.

———— 'Literature and Science under the Microscope', *Journal of Victorian Culture*, 11:2 (Autumn 2006), 301–15.

— 'Science and its Popularization', in Shattock (2010), 165–83.

Day, Lance and Ian McNeil, eds. *Biographical Dictionary of the History of Technology* (London: Routledge, 2003).

Day, Peter, ed. *The Philosopher's Tree: A Selection of Michael Faraday's Writings* (Bristol: Institute of Physics, 1999).

Delbourgo, James, *A Most Amazing Scene of Wonders: Electricity and Enlightenment in Early America* (Cambridge, MA: Harvard University Press, 2006).

Devine Jump, Harriet, *Nineteenth-Century Short Stories by Women: A Routledge Anthology* (London: Routledge, 1998).

Dickey, Gwyneth, 'Stanford Researchers Find Electrical Current Stemming from Plants', *Stanford University News Service* (13 April 2010) [http://news.

stanford.edu/pr/2010/pr-electric-current-plants-041310.html; accessed 22 November 2010].

Dreyfus, Hubert and Paul Rabinow, eds. *Michel Foucault: Beyond Structuralism and Hermeneutics* (Chicago: University of Chicago Press, 1982).

Eadie, Mervyn J. and Peter F. Bladin, *A Disease Once Sacred: A History of the Medical Understanding of Epilepsy* (Eastleigh: John Libbey, 2001).

Eckert, George M., Felix Gutmann and Hendrik Keyzer, *Electropharmacology* (Boca Raton: CRC Press, 1990).

Eliot, Simon, 'Some Trends in British Book Production 1800–1919', in Jordan and Patten (2003), 19–43.

Eliot, Simon and Jonathan Rose, eds. *A Companion to the History of the Book* (Malden, MA: Wiley-Blackwell, 2009).

Escott, T.H.S., 'Bulwer's Last Three Books', *Fraser's Magazine*, 9:54 (June 1874), 769.

Essig, Mark Regan, *Edison and the Electric Chair: A Story of Light and Death* (Stroud: Sutton, 2003).

Fara, Patricia, *Sympathetic Attractions: Magnetic Practices, Beliefs, and Symbolism in Eighteenth-Century England* (Princeton: Princeton University Press, 1996).

——— 'Alessandro Volta and the Politics of Pictures', *Endeavour*, 33:4 (2009), 126–7.

Feather, John, 'The British Book Market, 1600–1800', in Eliot and Rose (2009), 232–46.

Finkelman, Paul, *Encyclopedia of American Civil Liberties* 1 (New York: Routledge, 2006).

Fisch, Menachem and Simon Schaffer, eds. *William Whewell: A Composite Portrait* (Oxford: Clarendon, 1991).

Fleming, P.R., *A Short History of Cardiology* (Amsterdam: Rodopi, 1997).

Flichy, Patrice, *Understanding Innovation: A Socio-Technical Approach* (Cheltenham: Edward Elgar Publishing, 2007).

Flint, Kate, *The Victorians and the Visual Imagination* (Cambridge: Cambridge University Press, 2000).

Foucault, Michel, 'What is an Author?', tr. Donald F. Bouchard, in Bouchard and Simon (1977), 124–7.

Fox, Nichols, *Against the Machine: The Hidden Luddite Tradition in Literature, Art and Individual Lives* (Washington, DC: Island Press, 2002).

Frank, Alison Fleig, *Oil Empire: Visions of Prosperity in Austrian Galicia* (Cambridge, MA: Harvard University Press, 2005).

Friedman, Michael, *Kant and the Exact Sciences* (Cambridge, MA: Harvard University Press, 1992).

Fullmer, June Z., *Young Humphry Davy: The Making of an Experimental Chemist* (Philadelphia: American Philosophical Society, 2000).

Fyfe, Aileen, 'Conscientious Workmen or Booksellers' Hacks? The Professional Identities of Science Writers in the Mid-Nineteenth Century', *Isis*, 96:2 (June 2005), 192–223.

Fyfe, Aileen and Bernard Lightman, *Science in the Marketplace: Nineteenth-Century Sites and Experiences* (Chicago: University of Chicago Press, 2007).

Gage, John, *Color and Meaning: Art, Science, and Symbolism* (London: Thames & Hudson, 1999).

Galison, Peter, Stephen R. Graubard and Everett Mendelsohn, eds. *Science in Culture* (New Brunswick: Transaction Publishers, 2001).

Gentner, Dedre and D.R. Gentner, 'Flowing Waters or Teeming Crowds: Mental Models of Electricity', in Gentner and Stevens (1983), 99–129.

Gentner, Dedre and Albert L. Stevens, *Mental Models* (Hillsdale: Erlbaum, 1983).

Gersh, Meryl R., *Electrotherapy in Rehabilitation* (Philadelphia: Davis, 1992).

Gill, Richardson Benedict, *The Great Maya Droughts: Water, Life, and Death* (Albuquerque: University of New Mexico Press, 2001).

Gillespie, C.C., *The Edge of Objectivity: An Essay in the History of Scientific Ideas* (Princeton: Princeton University Press, 1960).

Gillespie, Gerald Ernest Paul, Manfred Engel and Bernard Dieterle, *Romantic Prose Fiction* (Amsterdam: John Benjamins, 2008).

Godfrey-Smith, Peter, 'Models and Fictions in Science', *Philosophical Studies*, 143:1 (March 2009), 101–16.

Gold, Barri J., *ThermoPoetics: Energy in Victorian Literature and Science* (Cambridge, MA: MIT Press, 2010).

Goldstein, Laurence, *The Flying Machine and Modern Literature* (Bloomington: Indiana University Press, 1986).

Golinski, Jan, *Science as Public Culture: Chemistry and Enlightenment in Britain, 1760–1820* (Cambridge: Cambridge University Press, 1992).

Gooday, Graeme, *The Morals of Measurement: Accuracy, Irony and Trust in Late Victorian Electrical Practice* (Cambridge: Cambridge University Press, 2004).

———— 'Profit and Prophecy: Electricity in the Late-Victorian Periodical', in Cantor et al. (2004), 238–54.

———— *Domesticating Electricity: Technology, Uncertainty and Gender, 1880–1914* (London: Pickering and Chatto, 2008).

Gooding, David, 'History in the Laboratory: Can We Tell What Really Went On?', in James (1989), 63–82.

———— '"Magnetic Curves" and the Magnetic Field: Experimentation and Representation in the History of a Theory', in Gooding et al. (1989), 183–223.

———— *Experiment and the Making of Meaning: Human Agency in Scientific Observation and Experiment* (Dordrecht: Kluwer Academic, 1990).

———— 'From Phenomenology to Field Theory: Faraday's Visual Reasoning', *Perspectives on Science* 14:1 (Spring 2006), 40–65.

Gooding, David, Trevor Pinch and Simon Schaffer, eds. *The Uses of Experiment: Studies in the Natural Sciences* (Cambridge: Cambridge University Press, 1989).

Grattan-Guinness, I., ed. *Landmark Writings in Western Mathematics 1640–1940* (Amsterdam: Elsevier B.V., 2005).

Griffin, Michael J. and Tom Moylan, eds. *Exploring the Utopian Impulse: Essays on Utopian Thought and Practice* (Bern: Peter Lang, 2007).

Hallion, Richard P., *Taking Flight: Inventing the Aerial Age from Antiquity through the First World War* (Oxford: Oxford University Press, 2003).

Hamilton, Susan, ed. *Animal Welfare and Anti-vivisection 1870–1910*, vol. 1 (London: Taylor & Francis, 2004).

Harman, Peter M., *Energy, Force, and Matter: The Conceptual Development of Nineteenth-Century Physics* (Cambridge: Cambridge University Press, 1982).

——— *The Natural Philosophy of James Clerk Maxwell* (Cambridge: Cambridge University Press, 1998).

Harris, Jason Marc, *Folklore and the Fantastic in Nineteenth-Century British Fiction* (Aldershot: Ashgate, 2008).

Hausman, William J., Peter Hertner and Mira Wilkins, *Global Electrification: Multinational Enterprise and International Finance in the History of Light and Power, 1878–2007* (Cambridge: Cambridge University Press, 2008).

Hawking, Stephen W., Untitled, *San Jose Mercury News*, Issue 0001240030 (23 January 2000).

——— *The Universe in a Nutshell* (London: Bantam, 2001).

Hayles, N. Katherine, *Chaos Bound: Orderly Disorder in Contemporary Literature and Science* (Ithaca, NY: Cornell University Press, 1990).

Hayward, Jennifer, *Consuming Pleasures: Active Audiences and Serial Fictions from Dickens to Soap Opera* (Lexington: University Press of Kentucky, 1997).

Heilbron, J.L. and James Bartholomew, *The Oxford Companion to the History of Modern Science* (Oxford: Oxford University Press, 2003).

Henson, Louise, "'In the Natural Course of Physical Things': Ghosts and Science in Charles Dickens's *All the Year Round*", in Henson et al. (2004), 113–23.

Henson, Louise, Geoffrey Cantor, Gowan Dawson, Richard Noakes, Sally Shuttleworth and Jonathan R. Topham, eds. *Culture and Science in the Nineteenth-Century Media* (Aldershot: Ashgate, 2004).

Himmelfarb, Gertrude, ed. *The Spirit of the Age: Victorian Essays* (New Haven: Yale University Press, 2007).

Hughes, Thomas P., *Networks of Power: Electrification in Western Society, 1880–1930* (Baltimore: Johns Hopkins University Press, 1983).

Huurdeman, Anton A., *The Worldwide History of Telecommunications* (London: Wiley-IEEE, 2003).

Irwin, Alan and Brian Wynne, *Misunderstanding Science? The Public Reconstruction of Science and Technology* (Cambridge: Cambridge University Press, 1996).

Iser, Wolfgang, *The Fictive and the Imaginary: Charting Literary Anthropology* (Baltimore: Johns Hopkins University Press, 1993).

Issitt, John, *Jeremiah Joyce: Radical, Dissenter and Writer* (Aldershot: Ashgate Publishing Ltd, 2006).

Jackson, Myles W., 'A Cultural History of Victorian Physical Science and Technology', *The Historical Journal*, 50 (2007), 253–64.

James, Frank A.J.L., ed. *The Development of the Laboratory: Essays on the Place of Experiment in Industrial Civilisation* (London: Macmillan, 1989).

——— 'Faraday, Michael (1791–1867)', *Oxford Dictionary of National Biography* (Oxford University Press, September 2004; online edn January 2008).

———, ed. *The Correspondence of Michael Faraday, 1855–1860*, vol. 5 (London: The Institution of Engineering and Technology, 2008).

——— *Michael Faraday: A Very Short Introduction* (Oxford: Oxford University Press, 2011).

James, J., *Light Microscopic Techniques in Biology and Medicine* (The Hague: Martinus Nijhoff, 1976).

Jenkins, Alice, *Space and the 'March of Mind': Literature and the Physical Sciences in Britain, 1815–1850* (Oxford: Oxford University Press, 2007).

——— *Michael Faraday's Mental Exercises: An Artisan Circle in Regency London* (Liverpool: Liverpool University Press, 2008).

Jones, Jason B., *Lost Causes: Historical Consciousness in Victorian Literature* (Columbus: Ohio State University Press, 2006).

Jordan, John O. and Robert L. Patten, eds. *Literature in the Marketplace: Nineteenth-Century British Publishing and Reading Practices* (Cambridge: Cambridge University Press, 2003).

Kaplan, Fred, 'The Mesmeric Mania: The Early Victorians and Animal Magnetism', *Journal of the History of Ideas*, 35:4 (October–December 1974), 691–702.

Keithley, Joseph F., *The Story of Electrical and Magnetic Measurements: from 500 B.C. to the 1940s* (New York: Institute of Electrical and Electronic Engineers, Inc., 1999).

Kent, Anthony, *Experimental Low Temperature Physics* (Basingstoke: Macmillan, 1993).

Kevles, Bettyann, *Naked to the Bone: Medical Imaging in the Twentieth Century* (New Brunswick: Rutgers University Press, 1997).

Killick, Tim, *British Short Fiction in the Early Nineteenth Century: The Rise of the Tale* (Aldershot: Ashgate, 2008).

King, Amy M., *Bloom: The Botanical Vernacular in the English Novel* (Oxford: Oxford University Press, 2003).

——— 'Searching out Science and Literature: Hybrid Narratives, New Methodological Directions, and Mary Russell Mitford's *Our Village*', *Literature Compass*, 4 (2007).

Kirby, M.W., *The Decline of British Economic Power Since 1870* (London: Routledge, 2006).

Knight, David M., 'Scientists and their Publics: Popularization of Science in the Nineteenth Century', in Nye (2003), 72–90.

——— 'Science and Professionalism in England, 1770–1830', *Proceedings of XIV International Congress of the History of Science, 1974*, vol. 1 (Tokyo, 1975) in Knight and Eddy (2005), 53–67.

────── *The Making of Modern Science: Science, Technology, Medicine and Modernity: 1789–1914* (Cambridge: Polity, 2009).

Knight, David M. and M.D. Eddy, eds. *Science and Beliefs: From Natural Philosophy to Natural Science, 1700–1900* (Aldershot: Ashgate 2005).

Knight, Mark and Emma Mason, *Nineteenth-Century Religion and Literature: An Introduction* (Oxford: Oxford University Press, 2006).

Kövecses, Zoltán, *Metaphor and Emotion: Language, Culture, and Body in Human Feeling* (Cambridge: Cambridge University Press, 2003).

────── *Metaphor in Culture: Universality and Variation* (Cambridge: Cambridge University Press, 2005).

Kuhn, Bernhard, *Autobiography and Natural Science in the Age of Romanticism: Rousseau, Goethe, Thoreau* (Farnham: Ashgate, 2009).

Kuhn, Thomas, *Scientific Revolutions* (Chicago: University of Chicago Press, 1996).

Lakoff, George and Mark Johnson, *Metaphors We Live By* (Chicago: University of Chicago Press, 1980).

────── *Philosophy in the Flesh: The Embodied Mind and its Challenge to Western Thought* (New York: Basic Books, 1999).

Lambert, Kevin, 'The Uses of Analogy: James Clerk Maxwell's "On Faraday's Lines of Force" and Early Victorian Analogical Argument', *British Journal for the History of Science*, 44:1 (March 2011), 61–88.

Latour, Bruno, *Science in Action* (Milton Keynes: Open University Press, 1987).

Law, Graham, *Serializing Fiction in the Victorian Press* (Basingstoke: Palgrave Macmillan, 2000).

Leplin, Jarrett, *A Novel Defense of Scientific Realism* (Oxford: Oxford University Press, 1997).

Levere, Trevor H., *Affinity and Matter: Elements of Chemical Philosophy 1800–1865* (Oxford: Clarendon Press, 1971).

Levine, George, ed. *One Culture: Essays in Science and Literature* (Madison: University of Wisconsin Press, 1987).

────── *Dying to Know: Scientific Epistemology and Narrative in Victorian England* (Chicago: University of Chicago Press, 2002).

Lewis, Albert C., 'Hamilton, Sir William Rowan (1805–1865)', *Oxford Dictionary of National Biography* (Oxford University Press, 2004) [www.oxforddnb.com/view/article/12148; accessed 25 January 2009].

Liddle, Dallas, *The Dynamics of Genre: Journalism and the Practice of Literature in Mid-Victorian Britain* (Charlottesville: University of Virginia Press, 2009).

Lightman, Bernard, *Victorian Popularizers of Science: Designing Nature for New Audiences* (Chicago: University of Chicago Press, 2007).

Lindley, David, *Uncertainty: Einstein, Heisenberg, Bohr, and the Struggle for the Soul of Science* (London: Doubleday Books, 2007).

Livingstone, David N. and Charles W.J. Withers, eds. *Geographies of Nineteenth-Century Science* (Chicago: University of Chicago Press, 2011).

Lyotard, Jean-François, *Phenomenology*, tr. Brian Beakley (Albany: SUNY Press, 2001).

McConnell, Anita, 'Lambe, John (1545/6–1628),' *Oxford Dictionary of National Biography* (Oxford University Press, 2004) [http://ezproxy.ouls.ox.ac.uk:2117/view/article/15925, accessed 25 October 2010].

McKitterick, David, ed. *The Cambridge History of the Book in Britain*, vol. 6, 1830–1914 (Cambridge: Cambridge University Press, 1987).

McNeil, Ian, ed. *An Encyclopaedia of the History of Technology* (London: Routledge, 1990).

Magnani, Lorenzo, Nancy J. Nersessian and Paul Thagard, *Model-Based Reasoning in Scientific Discovery* (New York: Kluwer Academic/Plenum, 1999).

Malcolm, Cheryl Alexander and David Malcolm, *A Companion to the British and Irish Short Story* (Oxford: Wiley-Blackwell, 2008).

Martin, J. and R. Harré, 'Metaphor in Science', in Miall (1982), 89–105.

Marvin, Carolyn, *When Old Technologies were New: Thinking about Electric Communication in the Late Nineteenth Century* (Oxford: Oxford University Press, 1988).

Masters, Gilbert M., *Renewable and Efficient Electric Power Systems* (London: Wiley-IEEE, 2004).

Mathus, Jill, *Shock, Memory and the Unconscious in Victorian Fiction* (Cambridge: Cambridge University Press, 2009).

Maunder, Andrew, *The Facts on File Companion to the British Short Story* (New York: Infobase Facts on File, 2007).

May, Charles Edward, 'Why Short Stories are Essential and Why They Are Seldom Read', in Winther et al. (2004), 14–31.

May, Charles Edward and Frank Northen Magill, *Critical Survey of Short Fiction: Essays, Research Tools, Indexes* (1981; Salem Press, 2001).

Miall, D.S., ed. *Metaphor: Problems and Perspectives* (Brighton: Harvester Press, 1982).

Middleton, L.M., 'Lumley, Benjamin (1811/12–1875)', rev. John Rosselli, *Oxford Dictionary of National Biography* (Oxford University Press, 2004; online edition, May 2007) [http://ezproxy.ouls.ox.ac.uk:2117/view/article/17174; accessed 9 December 2010].

Morus, Iwan Rhys, *Frankenstein's Children: Electricity, Exhibition, and Experiment in Early-Nineteenth-Century London* (Princeton: Princeton University Press, 1998).

———— *Bodies/Machines* (Oxford: Berg, 2002).

———— *When Physics Became King* (Chicago: University of Chicago Press, 2005).

———— 'Bodily Disciplines and Disciplined Bodies: Instruments, Skills and Victorian Electrotherapeutics', *Social History of Medicine*, 19:2 (2006), 241–59.

———— 'The Two Cultures of Electricity: Between Entertainment and Edification in Victorian Science', *Science and Education*, 16:6 (June 2007), 593–602.

———— *Shocking Bodies: Life, Death and Electricity in Victorian England* (Stroud: History Press Ltd, 2011).

Murdoch, Dugald, *Niels Bohr's Philosophy of Physics* (Cambridge: Cambridge University Press, 1989).

Mussell, James, *Science, Time and Space in the Late Nineteenth-Century Periodical Press: Movable Types* (Aldershot: Ashgate, 2007).

Myers, Greg, 'Nineteenth-Century Popularizations of Thermodynamics and the Rhetoric of Social Prophecy', *Victorian Studies*, 29:1 (Autumn 1985), 35–66.

——— 'Science of Women and Children: The Dialogue of Popular Science in the Nineteenth Century', in Christie and Shuttleworth (1989), 171–200.

Nersessian, Nancy J., 'Model-Based Reasoning in Conceptual Change', in Magnani et al. (1999), 5–22.

Niven, W.D., ed. *The Scientific Papers of James Clerk Maxwell* (New York: Dover Publications, Inc., 1965).

Noakes, Richard, 'Science in Mid-Victorian Punch', *Endeavour*, 26:3 (September 2002), 92–6.

Norris, Herbert and Oswald Curtis, *Nineteenth-Century Costume and Fashion* (Mineola: Constable, 1998).

Nye, M.J., ed. *The Cambridge History of Science*, vol. 5: *Modern Physical and Mathematical Sciences* (Cambridge: Cambridge University Press, 2003).

Ochs, Sidney, *A History of Nerve Functions: From Animal Spirits to Molecular Mechanisms* (Cambridge: Cambridge University Press, 2004).

O'Connor, Ralph, *The Earth on Show: Fossils and the Poetics of Popular Science, 1802–1856* (Chicago: University of Chicago Press, 2007).

Olby, R.C., G.N. Cantor, J.R.R. Christie and M.J.S. Hodge, eds. *Companion to the History of Modern Science* (London: Routledge, 1990).

Otis, Laura, *Networking: Communicating with Bodies and Machines in the Nineteenth Century* (Ann Arbor: University of Michigan Press, 2001).

———, ed. *Literature and Science in the Nineteenth Century: An Anthology* (Oxford: Oxford University Press, 2002).

——— 'Science Surveys and Histories of Literature: Reflections on an Uneasy Kinship', *Isis*, 101:3 (September 2010), 570–77.

Palumbo-DeSimone, Christine, *Sharing Secrets: Nineteenth-Century Women's Relations in the Short Story* (Madison: Fairleigh Dickinson University Press; London: Associated University Presses, 2000).

Pancaldi, Giuliano, *Volta: Science and Culture in the Age of Enlightenment* (Princeton: Princeton University Press, 2005).

Paoli, Letizia, *Mafia Brotherhoods: Organized Crime, Italian Style* (Oxford: Oxford University Press, 2003).

Papanelopoulou, F., A. Nieto-Galàn and E. Perdiguero, eds. *Popularizing Science and Technology in the European Periphery, 1800–2000* (Aldershot: Ashgate, 2009).

Paradis, James and Thomas Postlewait, eds. *Victorian Science and Victorian Values: Literary Perspectives* (New Brunswick: Rutgers University Press, 1985).

Pearson, Alan, *Robert Hunt* (St Austell: Federation of Old Cornwall Societies, 1976).

Peitzman, S.J. 'Bright's Disease and Bright's Generation—Toward Exact Medicine at Guy's Hospital', *Bulletin of the History of Medicine*, 55:3 (1981), 307–21.

Pitcher, Edward W.R., *An Anthology of the Short Story in 18th and 19th Century America* (Lewiston and Lampeter: E. Mellen Press, 2000).

Porter, Theodore M. 'The Objective Self', Review of *Objectivity* by Lorraine Daston and Peter Galison, *Victorian Studies*, 50:4 (Summer 2008), 641–7.

Pratt-Smith, Stella. 'The Other Serpents: Deviance and Contagion in Arthur Conan Doyle's The Speckled Band', *Victorian Newsletter* (Spring 2008: 113), 54–66.

Psillos, Stathis, *Scientific Realism: How Science Tracks Truth* (London: Routledge, 1999).

Pykett, Lyn, 'Reading the Periodical Press: Text and Context', in Brake et al. (1990), 3–18.

Rauch, Alan, *Useful Knowledge: The Victorians, Morality, and the March of Intellect* (Durham, NC: Duke University Press, 2001).

———— 'Poetry and Science', in Cronin et al. (2002), 475–92.

Reed, David, *Figures of Thought: Mathematics and Mathematical Texts* (London: Routledge, 1995).

Reid, Robin Anne, *Women in Science Fiction and Fantasy: Overviews* (Westport: Greenwood Publishing Group, 2009).

Richardson, Angélique, *Love and Eugenics in the Late Nineteenth Century: Rational Reproduction and the New Woman* (Oxford: Oxford University Press, 2003).

Rifkin, Jeremy, *The Third Industrial Revolution: How Lateral Power Is Transforming Energy, the Economy, and the World* (New York: Palgrave Macmillan, 2011).

Ritvo, Harriet, 'The View from the Hills: Environment and Technology in Victorian Periodicals', in Henson et al. (2004), 165–88.

Roe, Nicholas, *Samuel Taylor Coleridge and the Sciences of Life* (Oxford: Oxford University Press, 2001).

Roger, Joseph, *Epileptic Syndromes in Infancy, Childhood and Adolescence, International Workshop on Childhood Epileptology* (London: John Libby, 1985).

Rousseau, G.S. and Roy Porter, eds. *The Ferment of Knowledge: Studies in the Historiography of Eighteenth-Century Science* (Cambridge: Cambridge University Press, 1980).

Rowbottom, Margaret and Charles Susskind, *Electricity and Medicine: History of their Interaction* (San Francisco: San Francisco Press, 1984).

Rubery, Matthew, 'Victorian Print Culture, Journalism and the Novel', *Literature Compass*, 7 (2010), 290–300.

Rudy, Jason R., *Electric Meters: Victorian Physiological Poetics* (Athens, OH: Ohio University Press, 2009).

Rupke, Nicolaas A., *Vivisection in Historical Perspective*, Wellcome Institute for the History of Medicine (London: Croom Helm, 1987).

Samuelson, David N., 'Modes of Extrapolation: The Formulas of Hard SF', *Science Fiction Studies* 20:2 (July 1993), 191–232.

Saslow, Wayne M., *Electricity, Magnetism, and Light* (Amsterdam: Academic Press Elsvier Science, 2002).

Schaffer, Simon, 'Scientific Discoveries and the End of Natural Philosophy', *Social Studies of Science*, 16 (1986), 387–420.

——— 'The History and Geography of the Intellectual World: Whewell's Politics of Language', in Fisch and Schaffer (1991), 201–31.

Schenkel, Elmar and Stefan Welz, eds. *Lost Worlds and Mad Elephants: Literature, Science and Technology, 1700–1990* (Leipzig: Galda and Wilch, 1999).

Schiffer, Michael Brian, Kacy L. Hollenback and Carrie L. Bell, *Draw the Lightning Down: Benjamin Franklin and Electrical Technology in the Age of Enlightenment* (Berkeley and London: University of California Press, 2003).

Schlossberg, David, *Infections of Leisure*, 3rd edn (Washington, DC: ASM Press, 2004).

Schweber, S.S., 'Scientists as Intellectuals: the Early Victorians', in Paradis and Postlewait (1985), 1–38.

Seamon, David and Arthur Zajonc, *Goethe's Way of Science: A Phenomenology of Nature* (New York: State University of New York Press, 1998).

Secord, Anne, 'Pressed into Service: Specimens, Space, and Seeing in Botanical Practice', in Livingstone and Withers (2011), 283–310.

Secord, James A., 'Science, Technology and Mathematics', in McKitterick (1987), 443–74.

——— 'Extraordinary Experiments: Electricity and the Creation of Life in Victorian England', in Gooding et al. (1989), 337–83.

——— *Victorian Sensation: The Extraordinary Publication, Reception, and Secret Authorship of Vestiges of the Natural History of Creation* (Chicago: University of Chicago Press, 2000).

——— 'Quick and Magical Shaper of Science', *Science*, 297:5587 (September 2002), 1648–9.

———, ed. *Collected Works of Mary Somerville*, vol. 1 (Bristol: Thoemmes Continuum, 2004).

——— 'Knowledge in Transit', *Isis*, 95 (2004), 654–72.

——— Review: The Electronic Harvest, *The British Journal for the History of Science*, 38:4 (December 2005), 463–7.

Seed, David, ed. *A Companion to Science Fiction* (Blackwell Publishing, 2005), Blackwell Reference online [www.blackwellreference.com].

Semino, Elena, *Metaphor in Discourse* (Cambridge: Cambridge University Press, 2008).

Shattock, Joanne, ed. *The Cambridge Companion to English Literature, 1830–1914* (Cambridge: Cambridge University Press, 2010).

Shattock, Joanne and Michael Wolff, eds. *The Victorian Periodical Press: Samplings and Soundings* (Leicester: Leicester University Press, 1982).

Shapin, Stephen, 'Social Uses of Science', in Rousseau and Porter (1980), 93–139.

Sheets-Pyenson, Susan, 'Popular Science Periodicals in Paris and London: The Emergence of a Low Scientific Culture, 1820–1875', *Annals of Science*, 42:6 (November 1985), 549–72.

Shiach, Morag, *Discourse on Popular Culture: Class, Gender and History in Cultural Analysis, 1730 to the Present* (Stanford: Stanford University Press, 1989).

Shinn, Terry and Richard Whitley, eds. *Expository Science: Forms and Functions of Popularisation* (Dordrecht: D. Reidel, 1985).

Shippey, Tom, 'Literary Gatekeepers and the Fabril Tradition', in Westfahl (2002), 7–24.

Shuttleworth, Sally, *George Eliot and Nineteenth-Century Science: The Make-Believe of a Beginning* (Cambridge: Cambridge University Press, 1984).

Simon, Linda, *Dark Light: Electricity and Anxiety from the Telegraph to the X-Ray* (Boston: Houghton Mifflin Harcourt, 2005).

Simpson, Thomas K., *Figures of Thought: A Literary Appreciation of Maxwell's Treatise on Electricity and Magnetism* (Santa Fe: Green Lion Press, 2006).

Slusser, George, 'The Origins of Science Fiction', in Seed (2005), 27–42.

Smith, Crosbie, *The Science of Energy: A Cultural History of Energy Physics in Victorian Britain* (London: Athlone, 1998).

Smith, Jonathan, *Charles Darwin and Victorian Visual Culture* (Cambridge: Cambridge University Press, 2006).

Smith, Lindsay, *Victorian Photography, Painting and Poetry: The Enigma of Visibility in Ruskin, Morris and the Pre-Raphaelites* (Cambridge: Cambridge University Press, 1995).

Sneader, Walter, *Drug Discovery: A History* (Chichester: John Wiley, 2005).

Snow, Charles Percy, *The Two Cultures and the Scientific Revolution* (Cambridge: Cambridge University Press, 1959).

Snow, Stephanie, *Operations without Pain: The Practice and Science of Anaesthesia in Victorian Britain* (Basingstoke: Palgrave Macmillan, 2005).

Sorell, Tom, *Scientism: Philosophy and the Infatuation with Science* (London: Routledge, 1991).

Sparks, Tabitha, *The Doctor in the Victorian Novel: Family Practices* (Farnham: Ashgate, 2009).

Stafford, Barbara Maria, 'Visualization of Knowledge', in Brewer and Porter (1993), 462–77.

Stanley, Autumn, *Mothers and Daughters of Invention: Notes for a Revised History of Technology* (Metuchen and London: Scarecrow, 1993).

Stanley, Matthew, *Huxley's Church and Maxwell's Demon: From Theistic Science to Naturalistic Science* (Chicago: University of Chicago Press, 2014).

Stefanowitsch, Anatol and Stefan Thomas Gries, *Corpus-Based Approaches to Metaphor and Metonymy* (Berlin: Walter de Gruyter, 2006).

Sumpter, Caroline, 'The Cheap Press and the Reading Crowd: Visualizing Mass Culture and Modernity, 1838–1910', *Media History*, 12:3 (December 2006), 233–52.

Surhone, Lambert M., Mariam T. Tennoe and Susan F. Henssonow, eds. *Robert Hunt (Scientist)* (Saarbrücken: VDM Verlag Dr. Mueller AG and Co. Kg, 2010).

Sussman, Herbert, *Victorian Technology: Invention, Innovation, and the Rise of the Machine* (Santa Barbara: ABC-CLIO, 2009).

Temkin, Owsei, *The Falling Sickness: A History of Epilepsy from the Greeks to the Beginnings of Modern Neurology* (Baltimore: Johns Hopkins University Press, 1994).

Thompson, G.R., ed. *Edgar Allan Poe: Essays and Reviews* (New York: The Library of America, 1984).

Thorne-Murphy, Leslee, 'Students Researching Victorian Short Fiction', *Academic Exchange Quarterly*, 10:1 (Spring 2006), 232–6.

Topham, Jonathan R., 'Scientific Publishing and the Reading of Science in Nineteenth-Century Britain: A Historiographical Survey and Guide to Sources', *Studies in History and Philosophy of Science*, part A, 31:4 (2000), 559–612.

——— 'Publishing "Popular Science" in Early Nineteenth-Century Britain', in Fyfe and Lightman (2007), 135–68.

——— 'Rethinking the History of Science Popularization/Popular Science', in Papanelopoulou et al. (2009), 1–20.

Trotter, David, *Cooking with Mud: The Idea of Mess in Nineteenth-Century Fiction* (Oxford: Oxford University Press, 2000).

Tucker, Susie I., *Enthusiasm: A Study in Semantic Change* (Cambridge: Cambridge University Press, 1972).

Turner, Joseph, 'Maxwell's Method of Physical Analogy', *British Journal for the Philosophy of Science*, 6:23 (November 1955), 226–38.

Turner, Martha A., *Mechanism and the Novel: Science in the Narrative Process* (Cambridge: Cambridge University Press, 1993).

Van Fraassen, Bas C., *The Scientific Image* (Oxford: Oxford University Press, 1980).

Van Sciver, Steven W., *Helium Cryogenics* (New York: Plenum, 1986).

Voigts-Virchow, Eckhart, 'Melancholy Elephants and Virgin Machines: Technological Imagery and Mechanical Lacunae from Industrial Novels to Scientific Romances', in Schenkel and Welz (1999), 141–62.

Warner, William B., 'The Elevation of the Novel in England: Hegemony and Literary History', *ELH*, 59:3 (Autumn 1992), 577–96.

Welch, Robert, ed. *The Concise Oxford Companion to Irish Literature* (Oxford: Oxford University Press, 2004).

Westfahl, Gary, ed. *Science Fiction, Canonization, Marginalization, and the Academy* (Santa Barbara: Greenwood Publishing Group, 2002).

Westphal, M., *Local Therapies for Glioma: Present Status and Future Developments* (Wein: Springer, 2003).

White, Hayden V., *The Content of Form: Narrative Discourse and Historical Representation* (Baltimore: Johns Hopkins University Press, 1987).

Whitley, Richard, 'Knowledge Producers and Knowledge Acquirers: Popularization and a Relation between Scientific Fields and their Publics', in Shinn and Whitley (1985), 3–28.

Whitworth, Michael, *Einstein's Wake: Relativity, Metaphor and Modernist Literature* (Oxford: Oxford University Press, 2001).

——— *Modernism* (Malden, MA: Blackwell, 2007).

Williams, Carolyn, '"Inhumanly brought back to life and misery": Mary Wollstonecraft, Frankenstein, and the Royal Humane Society', *Women's Writing, the Elizabethan to Victorian Period*, 8:2 (2001), 213–34.

Williams, L. Pearce, 'The Physical Sciences in the First Half of the Nineteenth Century: Problems and Sources', *History of Science*, 1 (1962), 1–15.

Willis, Martin, *Mesmerists, Monsters, and Machines: Science Fiction and the Cultures of Science in the Nineteenth Century* (Kent, OH: Kent State University Press, 2006).

Winther, Per, Jakob Lothe and Hans H. Skei, eds. *The Art of Brevity: Excursions in Short Fiction Theory and Analysis* (Columbia: University of South Carolina Press, 2004).

Wise, M. Norton, 'The Mutual Embrace of Electricity and Magnetism', *Science*, 203 (1979), 1310–18.

Wood, Paul, *Science and Dissent in England, 1688–1945* (Aldershot: Ashgate, 2004).

Yeazell, Ruth Bernard, *Sex, Politics, and Science in the Nineteenth-Century Novel* (Baltimore: Johns Hopkins University Press, 1990).

Yeo, Richard, 'Science and Intellectual Authority in Mid-Nineteenth-Century Britain: Robert Chambers and "Vestiges of the Natural History of Creation"', *Victorian Studies*, 28:1 (Autumn 1984), 5–31.

——— *Defining Science: William Whewell, Natural Knowledge and Public Debate in Early Victorian Britain* (Cambridge: Cambridge University Press, 1993).

Zipes, Jack David, *When Dreams Came True: Classical Fairy Tales and their Tradition*, 2nd edn (New York: Routledge, 2007).

Zuck, David, Review: 'Operations without Pain: The Practice and Science of Anaesthesia in Victorian Britain', *Reviews in History* [www.history.ac.uk/reviews/review/573; accessed 3 December 2010].

# Index

Page numbers in italic indicate figures, bold indicates tables and notes are referenced as 1n2 (page 1 note 2).

Milton Keynes UK
Ingram Content Group UK Ltd.
UKHW031132141024
449569UK00006B/256